# ESSENTIAL SKILLS IN MATHS

## *Answer Book*

### BOOK 4

## Nelson

**Graham Newman and Ron Bull**

**National Curriculum coverage**

| Book | 1 | 2 | 3 | 4 | 5 |
|---|---|---|---|---|---|
| Levels | 3–4 | 4–5 | 5–6 | 6–7 | 7–8 |

**Thomas Nelson and Sons Ltd**
Nelson House   Mayfield Road
Walton-on-Thames   Surrey
KT12 5PL   UK

**Thomas Nelson Australia**
102 Dodds Street
South Melbourne
Victoria 3205   Australia

**Nelson Canada**
1120 Birchmount Road
Scarborough   Ontario
M1K 5G4   Canada

© R. Bull, G. Newman 1996

First published by Thomas Nelson and Sons Ltd 1996

I(T)P   Thomas Nelson is an International Thomson Publishing Company.

I(T)P   is used under licence.

ISBN 0-17-431466-3
NPN 9 8 7 6 5 4 3 2 1

Printed in China

# Contents

## SHAPE, SPACE AND MEASURES

## HANDLING DATA

# Number

## 1 STATING THE VALUE OF A DIGIT WITHIN A DECIMAL

### Exercise 1A

**1** (a) 0.8, 0.002     (b) $\frac{8}{10}$, $\frac{2}{1000}$

**2** (a) 0.05, 0.0004     (b) $\frac{5}{100}$, $\frac{4}{10\,000}$

**3** (a) 0.008, 0.0007     (b) $\frac{8}{1000}$, $\frac{7}{10\,000}$

**4** (a) 0.08, 0.0009     (b) $\frac{8}{100}$, $\frac{9}{10\,000}$

**5** (a) 0.02, 0.0007     (b) $\frac{2}{100}$, $\frac{7}{10\,000}$

**6** (a) 0.007, 0.0006     (b) $\frac{7}{1000}$, $\frac{6}{10\,000}$

**7** (a) 0.007, 0.0002     (b) $\frac{7}{1000}$, $\frac{2}{10\,000}$

**8** (a) 0.05, 0.0004     (b) $\frac{5}{100}$, $\frac{4}{10\,000}$

**9** (a) 0.03, 0.0005     (b) $\frac{3}{100}$, $\frac{5}{10\,000}$

**10** (a) 0.007, 0.000 01     (b) $\frac{7}{1000}$, $\frac{1}{100\,000}$

**11** (a) 0.3, 0.04     (b) $\frac{3}{10}$, $\frac{4}{100}$

**12** (a) 0.001, 0.000 05     (b) $\frac{1}{1000}$, $\frac{5}{100\,000}$

**13** (a) 0.0005, 0.000 009     (b) $\frac{5}{10\,000}$, $\frac{9}{1\,000\,000}$

**14** (a) 0.003, 0.0003     (b) $\frac{3}{1000}$, $\frac{3}{10\,000}$

**15** (a) 0.008, 0.000 04     (b) $\frac{8}{1000}$, $\frac{4}{100\,000}$

**16** (a) 0.8, 0.007     (b) $\frac{8}{10}$, $\frac{7}{1000}$

**17** (a) 0.05, 0.0006     (b) $\frac{5}{100}$, $\frac{6}{10\,000}$

**18** (a) 0.2, 0.0001     (b) $\frac{2}{10}$, $\frac{1}{10\,000}$

**19** (a) 0.07, 0.004     (b) $\frac{7}{100}$, $\frac{4}{1000}$

**20** (a) 0.2, 0.08     (b) $\frac{2}{10}$, $\frac{8}{100}$

**21** (a) 0.006, 0.0001     (b) $\frac{6}{1000}$, $\frac{1}{10\,000}$

**22** (a) 0.01, 0.0002     (b) $\frac{1}{100}$, $\frac{2}{10\,000}$

**23** (a) 0.9, 0.007     (b) $\frac{9}{10}$, $\frac{7}{1000}$

**24** (a) 0.8, 0.0001     (b) $\frac{8}{10}$, $\frac{1}{10\,000}$

**25** (a) 0.004, 0.0007     (b) $\frac{4}{1000}$, $\frac{7}{10\,000}$

**26** (a) 0.5, 0.01     (b) $\frac{5}{10}$, $\frac{1}{100}$

**27** (a) 0.0008, 0.000 06     (b) $\frac{8}{10\,000}$, $\frac{6}{100\,000}$

**28** (a) 0.008, 0.0007     (b) $\frac{8}{1000}$, $\frac{7}{10\,000}$

**29** (a) 0.006, 0.0003     (b) $\frac{6}{1000}$, $\frac{3}{10\,000}$

**30** (a) 0.005, 0.000 01     (b) $\frac{5}{1000}$, $\frac{1}{100\,000}$

### Exercise 1B

**1** (a) 0.07, 0.0004     (b) $\frac{7}{100}$, $\frac{4}{10\,000}$

**2** (a) 0.8, 0.009     (b) $\frac{8}{10}$, $\frac{9}{1000}$

**3** (a) 0.005, 0.0004     (b) $\frac{5}{1000}$, $\frac{4}{10\,000}$

**4** (a) 0.09, 0.0008     (b) $\frac{9}{100}$, $\frac{8}{10\,000}$

**5** (a) 0.7, 0.0006     (b) $\frac{7}{10}$, $\frac{6}{10\,000}$

**6** (a) 0.09, 0.001     (b) $\frac{9}{100}$, $\frac{1}{1000}$

**7** (a) 0.001, 0.000 02     (b) $\frac{1}{1000}$, $\frac{2}{100\,000}$

**8** (a) 0.0002, 0.000 01     (b) $\frac{2}{10\,000}$, $\frac{1}{100\,000}$

**9** (a) 0.6, 0.005     (b) $\frac{6}{10}$, $\frac{5}{1000}$

**10** (a) 0.0006, 0.000 08     (b) $\frac{6}{10\,000}$, $\frac{8}{100\,000}$

**11** (a) 0.03, 0.0001     (b) $\frac{3}{100}$, $\frac{1}{10\,000}$

**12** (a) 0.009, 0.0001     (b) $\frac{9}{1000}$, $\frac{1}{10\,000}$

**13** (a) 0.06, 0.0002     (b) $\frac{6}{100}$, $\frac{2}{10\,000}$

**14** (a) 0.006, 0.000 005     (b) $\frac{6}{1000}$, $\frac{5}{1\,000\,000}$

**15** (a) 0.0001, 0.000 003     (b) $\frac{1}{10\,000}$, $\frac{3}{1\,000\,000}$

**16** (a) 0.3, 0.0002     (b) $\frac{3}{10}$, $\frac{2}{10\,000}$

**17** (a) 0.1, 0.09     (b) $\frac{1}{10}$, $\frac{9}{100}$

**18** (a) 0.002, 0.0008     (b) $\frac{2}{1000}$, $\frac{8}{10\,000}$

**19** (a) 0.6, 0.009     (b) $\frac{6}{10}$, $\frac{9}{1000}$

**20** (a) 0.3, 0.0002     (b) $\frac{3}{10}$, $\frac{2}{10\,000}$

**21** (a) 0.09, 0.008     (b) $\frac{9}{100}$, $\frac{8}{1000}$

**22** (a) 0.01, 0.0004     (b) $\frac{1}{100}$, $\frac{4}{10\,000}$

**23** (a) 0.7, 0.007     (b) $\frac{7}{10}$, $\frac{7}{1000}$

**24** (a) 0.0008, 0.000 06     (b) $\frac{8}{10\,000}$, $\frac{6}{100\,000}$

**25** (a) 0.0001, 0.000 003     (b) $\frac{1}{10\,000}$, $\frac{3}{1\,000\,000}$

**26** (a) 0.09, 0.0007     (b) $\frac{9}{100}$, $\frac{7}{10\,000}$

**27** (a) 0.02, 0.0009     (b) $\frac{2}{100}$, $\frac{9}{10\,000}$

**28** (a) 0.0005, 0.000 004     (b) $\frac{5}{10\,000}$, $\frac{4}{1\,000\,000}$

**29** (a) 0.08, 0.001     (b) $\frac{8}{100}$, $\frac{1}{1000}$

**30** (a) 0.03, 0.000 04     (b) $\frac{3}{100}$, $\frac{4}{100\,000}$

# 2 MENTAL MULTIPLICATION AND DIVISION

## Exercise 2A

| | | |
|---|---|---|
| **1** 520 | **2** 26 100 | **3** 820 |
| **4** 1.23 | **5** 18.7 | **6** 1.19 |
| **7** 34 400 | **8** 0.25 | **9** 348 000 |
| **10** 25.6 | **11** 0.301 | **12** 19 000 |
| **13** 70.9 | **14** 8.19 | **15** 22.9 |
| **16** 2100 | **17** 549 000 | **18** 0.363 |
| **19** 41.6 | **20** 2.83 | **21** 98 000 |
| **22** 48 300 | **23** 3000 | **24** 88.9 |
| **25** 2128 000 | **26** 6.48 | **27** 67 300 |
| **28** 8640 | **29** 0.184 | **30** 764 000 |

## Exercise 2B

| | | |
|---|---|---|
| **1** 3300 | **2** 25 200 | **3** 13.2 |
| **4** 162 000 | **5** 29 | **6** 39 900 |
| **7** 46 000 | **8** 25.3 | **9** 1.12 |
| **10** 0.248 | **11** 19 200 | **12** 2.2 |
| **13** 58.3 | **14** 2.45 | **15** 0.62 |
| **16** 20.2 | **17** 2940 | **18** 95 000 |
| **19** 11 480 | **20** 0.612 | **21** 39.4 |
| **22** 80 700 | **23** 630 400 | **24** 83.1 |
| **25** 8.37 | **26** 19 320 | **27** 492 800 |
| **28** 2.24 | **29** 300 000 | **30** 864 000 |

# 3 ORDERING DECIMALS

## Exercise 3A

**1** 0.63, 0.64, 0.69
**2** 0.19, 0.33, 0.72
**3** 0.1, 0.3, 0.6, 0.8
**4** 0.4, 0.5, 0.8
**5** 0.222, 0.345, 0.949, 0.955
**6** 0.3024, 0.4145, 0.8050, 0.9247
**7** 0.035, 0.267, 0.305
**8** 0.003, 0.0333, 0.3
**9** 0.006, 0.06, 0.6
**10** 0.034, 0.044, 0.059
**11** 4.008, 4.048, 4.208
**12** 0.044, 0.064, 0.204
**13** 0.09, 0.9, 0.99, 9.9
**14** 0.34, 0.43, 3.4, 4.3
**15** 34.234, 34.432, 34.442
**16** 3.0048, 3.0078, 3.008
**17** 0.007 77, 0.0307, 0.0377
**18** 5.055, 5.505, 5.550
**19** 2.07, 8.07, 8.7, 20.7
**20** 0.567, 0.576, 0.583, 0.585
**21** 8.099, 8.909, 8.999
**22** 3.552, 3.652, 4.553, 4.558
**23** 1.001, 1.1, 1.101, 1.11
**24** 0.002, 0.02, 0.2, 2.0
**25** 0.04, 0.86, 4.0, 5.4
**26** 5.43, 5.53, 6.03, 6.63
**27** 7.202, 7.204, 7.305, 7.307
**28** 0.235, 0.727, 0.77, 2.35
**29** 0.033, 0.330, 3.0, 3.03
**30** 9.90, 9.93, 99.1, 99.5

## Exercise 3B

**1** 0.003, 0.03, 0.3
**2** 0.4, 0.6, 0.8
**3** 0.022, 0.042, 0.202
**4** 0.3, 0.5, 0.7, 0.8
**5** 0.024, 0.044, 0.049
**6** 0.29, 0.43, 0.92
**7** 0.51, 0.52, 0.59
**8** 0.333, 0.433, 0.684, 0.941
**9** 7.009, 7.049, 7.109
**10** 0.023, 0.243, 0.343
**11** 0.009, 0.0999, 0.9
**12** 14.124, 14.421, 14.441
**13** 5.011, 5.101, 5.111
**14** 0.2053, 0.3254, 0.9143, 0.9242
**15** 7.0069, 7.0089, 7.009
**16** 8.088, 8.808, 8.880
**17** 0.34, 0.43, 3.4, 4.3
**18** 0.07, 0.7, 0.77, 7.7
**19** 0.05, 0.85, 5.0, 7.7
**20** 342.49, 342.53, 343.02
**21** 0.425, 0.825, 0.88, 1.85
**22** 0.008 88, 0.0108, 0.0188
**23** 6.303, 6.304, 6.403, 6.506
**24** 0.325, 0.352, 0.363, 0.365
**25** 0.022, 0.220, 2.0, 2.02
**26** 4.04, 4.07, 4.4, 5.4
**27** 0.006, 0.06, 0.6, 6.0
**28** 6.734, 6.865, 7.665, 7.667
**29** 9.714, 9.9, 9.908, 9.99
**30** 2.20, 2.23, 22.2, 22.4

# 4 EVALUATING POWERS

## Exercise 4A

| | | |
|---|---|---|
| **1** 36 | **2** 64 | **3** 64 |
| **4** 4096 | **5** 512 | **6** 256 |
| **7** 625 | **8** 729 | **9** 256 |

10 49    11 7776    12 4096

**10** 49 **11** 7776 **12** 4096
**13** 10 000 **14** 14 641 **15** 2197
**16** 2 **17** 7 **18** 3
**19** 3 **20** 5 **21** 6
**22** 675 **23** 1296 **24** 966
**25** 189 **26** 371 **27** 121 000
**28** 1 327 104 **29** 746 496 **30** 8281

## Exercise 4B

**1** 9 **2** 1296 **3** 81
**4** 81 **5** 16 **6** 512
**7** 2187 **8** 3125 **9** 2401
**10** 6561 **11** 100 **12** 1331
**13** 225 **14** 1728 **15** 625
**16** 3 **17** 2 **18** 3
**19** 3 **20** 3 **21** 10
**22** 50 625 **23** 73 728 **24** 2048
**25** 88 **26** 531 441 **27** 16 384
**28** 7825 **29** 104 **30** 559 872

## 5 EXPRESSING POSITIVE INTEGERS AS PRODUCTS OF PRIMES

## Exercise 5A

**1** $2 \times 2 \times 2$    **2** $3 \times 5 \times 7$
**3** $2 \times 3 \times 3$ **4** $5 \times 5 \times 7$
**5** $2 \times 5 \times 5$ **6** $2 \times 2 \times 3$
**7** $3 \times 3 \times 3 \times 5$ **8** $2 \times 7 \times 7$
**9** $2 \times 3 \times 5$ **10** $7 \times 7 \times 11$
**11** $2 \times 2 \times 5$ **12** $3 \times 3 \times 17$
**13** $2 \times 3 \times 5 \times 7$ **14** $2 \times 11 \times 11$
**15** $2 \times 2 \times 2 \times 2 \times 7$ **16** $11 \times 11$
**17** $3 \times 11 \times 19$ **18** $3 \times 3 \times 7$
**19** $2 \times 2 \times 2 \times 3 \times 3$ **20** $2 \times 2 \times 11$
**21** $2 \times 2 \times 3 \times 3$ **22** $3 \times 5 \times 5$
**23** $2 \times 3 \times 3 \times 3 \times 3$ **24** $2 \times 3 \times 13$
**25** $2 \times 2 \times 2 \times 3$ **26** $2 \times 5 \times 13$
**27** $2 \times 2 \times 3 \times 5 \times 7 \times 7$ **28** $2 \times 7 \times 31$
**29** $5 \times 5 \times 5 \times 7$ **30** $2 \times 2 \times 2 \times 5 \times 5$

## Exercise 5B

**1** $2 \times 2 \times 3$    **2** $2 \times 7 \times 11$
**3** $2 \times 3 \times 7$ **4** $2 \times 2 \times 3 \times 5$
**5** $2 \times 5 \times 7$ **6** $2 \times 2 \times 7$
**7** $3 \times 5 \times 11$ **8** $2 \times 2 \times 2 \times 5$
**9** $3 \times 3 \times 5$ **10** $2 \times 2 \times 13$
**11** $2 \times 3 \times 11$ **12** $2 \times 11 \times 13$
**13** $2 \times 3 \times 5 \times 11$ **14** $2 \times 5 \times 19$
**15** $3 \times 3 \times 13$ **16** $5 \times 5 \times 11$
**17** $2 \times 2 \times 2 \times 2 \times 5$ **18** $2 \times 2 \times 17$
**19** $3 \times 5 \times 19$ **20** $3 \times 11 \times 11$

**21** $2 \times 3 \times 31$ **22** $2 \times 2 \times 3 \times 11$
**23** $2 \times 2 \times 5 \times 5 \times 7 \times 11$ **24** $2 \times 7$
**25** $3 \times 3 \times 3 \times 11$ **26** $2 \times 2 \times 31$
**27** $13 \times 13$ **28** $2 \times 2 \times 2 \times 2 \times 11$
**29** $5 \times 11 \times 11 \times 13$
**30** $3 \times 3 \times 3 \times 3 \times 3 \times 3$

## 6 ROUNDING TO DECIMAL PLACES

## Exercise 6A

**1** (a) 0.9    (b) 0.92    (c) 0.921
**2** (a) 0.6 (b) 0.64 (c) 0.640
**3** (a) 0.5 (b) 0.52 (c) 0.523
**4** (a) 0.4 (b) 0.44 (c) 0.443
**5** (a) 5.9 (b) 5.90 (c) 5.896
**6** (a) 0.0 (b) 0.03 (c) 0.032
**7** (a) 7.5 (b) 7.52 (c) 7.525
**8** (a) 0.6 (b) 0.62 (c) 0.619
**9** (a) 1.0 (b) 0.99 (c) 0.986
**10** (a) 9.1 (b) 9.08 (c) 9.081
**11** (a) 6.0 (b) 5.97 (c) 5.966
**12** (a) 3.1 (b) 3.11 (c) 3.111
**13** (a) 8.2 (b) 8.18 (c) 8.177
**14** (a) 0.3 (b) 0.27 (c) 0.269
**15** (a) 0.1 (b) 0.13 (c) 0.134
**16** (a) 8.7 (b) 8.71 (c) 8.706
**17** (a) 12.1 (b) 12.11 (c) 12.114
**18** (a) 5.4 (b) 5.35 (c) 5.352
**19** (a) 4.7 (b) 4.69 (c) 4.691
**20** (a) 12.1 (b) 12.11 (c) 12.105
**21** (a) 7.0 (b) 7.05 (c) 7.046
**22** (a) 7.7 (b) 7.71 (c) 7.714
**23** (a) 2.7 (b) 2.74 (c) 2.745
**24** (a) 0.4 (b) 0.40 (c) 0.403
**25** (a) 81.8 (b) 81.79 (c) 81.793
**26** (a) 12.2 (b) 12.22 (c) 12.222
**27** (a) 7.6 (b) 7.62 (c) 7.619
**28** (a) 0.3 (b) 0.27 (c) 0.274
**29** (a) 10.3 (b) 10.28 (c) 10.282
**30** (a) 34.2 (b) 34.15 (c) 34.155

## Exercise 6B

**1** (a) 0.0    (b) 0.01    (c) 0.015
**2** (a) 0.5 (b) 0.47 (c) 0.469
**3** (a) 0.8 (b) 0.79 (c) 0.791
**4** (a) 0.6 (b) 0.64 (c) 0.644
**5** (a) 1.0 (b) 0.99 (c) 0.994
**6** (a) 2.1 (b) 2.08 (c) 2.077
**7** (a) 0.8 (b) 0.80 (c) 0.802
**8** (a) 0.9 (b) 0.87 (c) 0.870

| 9 | (a) 5.9 | (b) 5.92 | (c) 5.924 |
|---|---|---|---|
| 10 | (a) 0.1 | (b) 0.10 | (c) 0.101 |
| 11 | (a) 1.3 | (b) 1.25 | (c) 1.254 |
| 12 | (a) 7.3 | (b) 7.28 | (c) 7.276 |
| 13 | (a) 1.0 | (b) 0.96 | (c) 0.960 |
| 14 | (a) 4.8 | (b) 4.80 | (c) 4.803 |
| 15 | (a) 0.9 | (b) 0.89 | (c) 0.893 |
| 16 | (a) 0.4 | (b) 0.39 | (c) 0.391 |
| 17 | (a) 8.0 | (b) 8.05 | (c) 8.048 |
| 18 | (a) 0.3 | (b) 0.31 | (c) 0.312 |
| 19 | (a) 8.3 | (b) 8.34 | (c) 8.343 |
| 20 | (a) 10.1 | (b) 10.09 | (c) 10.095 |
| 21 | (a) 12.9 | (b) 12.94 | (c) 12.939 |
| 22 | (a) 13.8 | (b) 13.78 | (d) 13.780 |
| 23 | (a) 9.7 | (b) 9.74 | (c) 9.742 |
| 24 | (a) 13.8 | (b) 13.75 | (c) 13.753 |
| 25 | (a) 0.3 | (b) 0.26 | (c) 0.259 |
| 26 | (a) 0.9 | (b) 0.88 | (c) 0.877 |
| 27 | (a) 0.0 | (b) 0.00 | (c) 0.003 |
| 28 | (a) 0.4 | (b) 0.35 | (c) 0.355 |
| 29 | (a) 9.3 | (b) 9.30 | (c) 9.296 |
| 30 | (a) 7.2 | (b) 7.24 | (c) 7.241 |

# 7 USING A CALCULATOR FOR CALCULATIONS

## Exercise 7A

Answers are written correct to 3 decimal places.

| 1 | 129.132 | 2 | 3.250 | 3 | 2020.155 |
|---|---|---|---|---|---|
| 4 | 17.316 | 5 | 49.902 | 6 | 1.678 |
| 7 | 31.753 | 8 | 1.477 | 9 | 183.175 |
| 10 | 18.985 | 11 | 2.980 | 12 | 2.552 |
| 13 | 25.109 | 14 | 31.643 | 15 | 4.142 |
| 16 | 249.119 | 17 | 23.49 | 18 | 37.661 |
| 19 | 4.326 | 20 | 0.061 | 21 | 15.071 |
| 22 | 15.698 | 23 | 1.213 | 24 | 62.053 |
| 25 | 0.188 | 26 | 0.403 | 27 | 90.229 |
| 28 | 1.416 | 29 | 4.989 | 30 | 70.141 |

## Exercise 7B

Answers are written correct to 3 decimal places.

| 1 | 20.794 | 2 | 11.989 | 3 | 2.462 |
|---|---|---|---|---|---|
| 4 | 31.007 | 5 | 1.328 | 6 | 0.063 |
| 7 | 3710.646 | 8 | 11.416 | 9 | 1.737 |
| 10 | 70.375 | 11 | 2929.203 | 12 | 136.402 |
| 13 | 28.803 | 14 | 0.432 | 15 | 0.239 |
| 16 | 1.382 | 17 | 3.380 | 18 | 98.631 |
| 19 | 2.784 | 20 | 0.042 | 21 | 5896.620 |
| 22 | 219.859 | 23 | 1.515 | 24 | 16.873 |
| 25 | 0.146 | 26 | 5.397 | 27 | 0.141 |
| 28 | 1424.140 | 29 | 0.197 | 30 | 19.667 |

# 8 ROUNDING TO SIGNIFICANT FIGURES

## Exercise 8A

| 1 | (a) 0.7 | (b) 0.70 | (c) 0.695 |
|---|---|---|---|
| 2 | (a) 0.8 | (b) 0.77 | (c) 0.774 |
| 3 | (a) 0.5 | (b) 0.53 | (c) 0.531 |
| 4 | (a) 0.6 | (b) 0.56 | (c) 0.555 |
| 5 | (a) 0.9 | (b) 0.90 | (c) 0.901 |
| 6 | (a) 0.6 | (b) 0.60 | (c) 0.600 |
| 7 | (a) 3 | (b) 3.3 | (c) 3.30 |
| 8 | (a) 5 | (b) 4.9 | (c) 4.86 |
| 9 | (a) 8 | (b) 8.2 | (c) 8.23 |
| 10 | (a) 5 | (b) 5.5 | (c) 5.46 |
| 11 | (a) 9 | (b) 9.4 | (c) 9.43 |
| 12 | (a) 8 | (b) 8.3 | (c) 8.26 |
| 13 | (a) 50 | (b) 50 | (c) 49.6 |
| 14 | (a) 500 | (b) 550 | (c) 549 |
| 15 | (a) 1 | (b) 1.3 | (c) 1.35 |
| 16 | (a) 20 | (b) 22 | (c) 21.5 |
| 17 | (a) 100 | (b) 130 | (c) 125 |
| 18 | (a) 8000 | (b) 8000 | (c) 7970 |
| 19 | (a) 400 | (b) 400 | (c) 404 |
| 20 | (a) 0.05 | (b) 0.055 | (c) 0.0548 |
| 21 | (a) 9 | (b) 9.2 | (c) 9.21 |
| 22 | (a) 200 | (b) 190 | (c) 191 |
| 23 | (a) 0.1 | (b) 0.10 | (c) 0.104 |
| 24 | (a) 500 | (b) 470 | (c) 474 |
| 25 | (a) 8000 | (b) 8100 | (c) 8100 |
| 26 | (a) 400 | (b) 390 | (c) 391 |
| 27 | (a) 100 | (b) 98 | (c) 98.1 |
| 28 | (a) 0.8 | (b) 0.83 | (c) 0.835 |
| 29 | (a) 600 | (b) 630 | (c) 629 |
| 30 | (a) 10 | (b) 10 | (c) 10.2 |

## Exercise 8B

| 1 | (a) 1 | (b) 0.98 | (c) 0.984 |
|---|---|---|---|
| 2 | (a) 0.2 | (b) 0.20 | (c) 0.196 |
| 3 | (a) 0.9 | (b) 0.86 | (c) 0.857 |
| 4 | (a) 0.5 | (b) 0.47 | (c) 0.473 |
| 5 | (a) 0.3 | (b) 0.28 | (c) 0.277 |
| 6 | (a) 3 | (b) 2.9 | (c) 2.88 |
| 7 | (a) 8 | (b) 7.8 | (c) 7.81 |
| 8 | (a) 30 | (b) 28 | (c) 28.0 |
| 9 | (a) 60 | (b) 58 | (c) 57.8 |
| 10 | (a) 900 | (b) 880 | (c) 878 |
| 11 | (a) 8 | (b) 8.1 | (c) 8.10 |
| 12 | (a) 700 | (b) 700 | (c) 697 |
| 13 | (a) 4000 | (b) 4300 | (c) 4320 |
| 14 | (a) 20 | (b) 22 | (c) 21.9 |
| 15 | (a) 900 | (b) 940 | (c) 935 |

**16** (a) 4000    (b) 3800    (c) 3830
**17** (a) 6    (b) 6.1    (c) 6.15
**18** (a) 0.7    (b) 0.68    (c) 0.678
**19** (a) 700    (b) 740    (c) 741
**20** (a) 6000    (b) 5500    (c) 5530
**21** (a) 0.03    (b) 0.026    (c) 0.0261
**22** (a) 200    (b) 180    (c) 184
**23** (a) 9000    (b) 8500    (c) 8500
**24** (a) 50    (b) 54    (c) 54.4
**25** (a) 700    (b) 740    (c) 740
**26** (a) 200    (b) 200    (c) 200
**27** (a) 2    (b) 2.1    (c) 2.11
**28** (a) 400    (b) 410    (c) 414
**29** (a) 400    (b) 430    (c) 426
**30** (a) 2000    (b) 2100    (c) 2060

# 9 ESTIMATES FOR CALCULATIONS

## Exercise 9A

**1** 60    **2** 40    **3** 20    **4** 810
**5** 50    **6** 160    **7** 45    **8** 400
**9** 25    **10** $\frac{1}{2}$    **11** 50    **12** $\frac{1}{12}$
**13** 60    **14** 200    **15** 160    **16** 750
**17** 1000    **18** 750    **19** 4800    **20** 80
**21** 200    **22** 80    **23** 30    **24** 100
**25** 500    **26** 437.5    **27** 210    **28** 8
**29** $6\frac{1}{4}$    **30** 100

## Exercise 9B

**1** 80    **2** 5000    **3** 20    **4** $\frac{1}{4}$
**5** $1\frac{1}{3}$    **6** 20    **7** 150    **8** $\frac{4}{5}$
**9** 12    **10** 100    **11** 40    **12** 8
**13** 200    **14** $2\frac{1}{2}$    **15** 2    **16** $\frac{3}{8}$
**17** 150    **18** 4    **19** 100    **20** 16
**21** 800    **22** 3    **23** 100    **24** $33\frac{1}{3}$
**25** 300    **26** 14    **27** 20    **28** 100
**29** 3750    **30** $1\frac{7}{8}$

# 10 MEASUREMENT TO HALF A UNIT

## Exercise 10A

**1** (a) 40.5 cm    (b) 41.5 cm
**2** (a) 130.5 km    (b) 131.5 km
**3** (a) 94.5 m    (b) 95.5 m
**4** (a) 2.375 cm    (b) 2.385 cm
**5** (a) 10.75 t    (b) 10.85 t
**6** (a) 0.4135 m    (b) 0.4145 m
**7** (a) 43.15 cm    (b) 43.25 cm
**8** (a) 0.3845 *l*    (b) 0.3855 *l*
**9** (a) 31.5 cm    (b) 32.5 cm
**10** (a) 2.4435 m    (b) 2.4445 m
**11** (a) 56.45 cl    (b) 56.55 cl
**12** (a) 7.2785 kg    (b) 7.2795 kg
**13** (a) 0.535 cm    (b) 0.545 cm
**14** (a) 3.085 g    (b) 3.095 g
**15** (a) 99.5 m.p.h.    (b) 100.5 m.p.h.
**16** (a) 3.2875 cm    (b) 3.2885 cm
**17** (a) 2.225 g    (b) 2.235 g
**18** (a) 21.25 cm    (b) 21.25 cm
**19** (a) 56.5 m    (b) 57.5 m
**20** (a) 5.485 *l*    (b) 5.495 *l*
**21** (a) 23.55 kg    (b) 23.65 kg
**22** (a) 2.685 ml    (b) 2.695 ml
**23** (a) 5.3155 cm    (b) 5.3165 cm
**24** (a) 4.765 s    (b) 4.775 s
**25** (a) 92.5 km    (b) 93.5 km
**26** (a) 14.2925 cm    (b) 14.2935 cm
**27** (a) 16.75 t    (b) 16.85 t
**28** (a) 0.2145 m    (b) 0.2155 m
**29** (a) 1.105 t    (b) 1.115 t
**30** (a) 4.1475 km    (b) 4.1485 km

## Exercise 10B

**1** (a) 22.5 ml    (b) 23.5 ml
**2** (a) 46.5 g    (b) 47.5 g
**3** (a) 236.5 t    (b) 237.5 t
**4** (a) 1.725 kg    (b) 1.735 kg
**5** (a) 7.945 m    (b) 7.955 m
**6** (a) 20.05 g    (b) 20.15 g
**7** (a) 5.2445 t    (b) 5.2455 t
**8** (a) 15.75 km    (b) 15.85 km
**9** (a) 7.3805 *l*    (b) 7.3815 *l*
**10** (a) 17.15 m    (b) 17.25 m
**11** (a) 5.1075 cm    (b) 5.1085 cm
**12** (a) 9.5 m.p.h.    (b) 10.5 m.p.h.
**13** (a) 0.6805 cl    (b) 0.6815 cl
**14** (a) 37.55 s    (b) 37.65 s
**15** (a) 7.2215 m    (b) 7.2225 m
**16** (a) 0.0005 *l*    (b) 0.0015 *l*
**17** (a) 0.6675 cm    (b) 0.6685 cm
**18** (a) 21.5 m    (b) 22.5 m
**19** (a) 0.035 *l*    (b) 0.045 *l*
**20** (a) 30.55 s    (b) 30.65 s
**21** (a) 3.075 cm    (b) 3.085 cm
**22** (a) 33.3305 kg    (b) 3.3315 kg
**23** (a) 2.105 km    (b) 2.115 km
**24** (a) 2.065 *l*    (b) 2.075 *l*

**25** (a) 1.20045 m    (b) 1.20055 m
**26** (a) 83.5 cl    (b) 84.5 cl
**27** (a) 1.1275 kg    (b) 1.1285 kg
**28** (a) 1.215 s    (b) 1.225 s
**29** (a) 2.60045 t    (b) 2.60055 t
**30** (a) 3.655 g    (b) 3.665 g

# 11   RATIO

## Exercise 11A

| | | |
|---|---|---|
| **1** 1 : 4 | **2** 2 : 3 | **3** 1 : 8 |
| **4** 1 : 4 | **5** 2 : 3 | **6** 4 : 5 |
| **7** 3 : 2 | **8** 83 : 43 | **9** 3 : 2 |
| **10** 1 : 5 | **11** 7 : 5 | **12** 3 : 20 |
| **13** 3 : 14 | **14** 1 : 4 | **15** 29 : 60 |
| **16** 4 : 9 | **17** 3 : 5 | **18** 1 : 50 : 2 |
| **19** 13 : 5 : 8 | **20** 5 : 2 : 1000 | **21** 1 : 6 |
| **22** 24 : 1 | **23** 1 : 0.25 | **24** 1 : 5 |
| **25** 1 : 0.5 | **26** 1 : 2.5 | **27** 1 : 1.2 |
| **28** 1 : 2.2 | **29** $1 : 1\frac{1}{3}$ | **30** $1 : 1\frac{1}{7}$ |

## Exercise 11B

| | | |
|---|---|---|
| **1** 1 : 2 | **2** 2 : 3 | **3** 5 : 2 |
| **4** 6 : 1 | **5** 4 : 9 | **6** 2 : 3 |
| **7** 5 : 7 | **8** 3 : 4 | **9** 1 : 20 |
| **10** 3 : 10 | **11** 4 : 1 | **12** 2 : 7 |
| **13** 1 : 2 | **14** 3 : 8 | **15** 2 : 3 |
| **16** 3 : 2 | **17** 5 : 4 | **18** 3 : 8 : 2 |
| **19** 4 : 5 : 9 | **20** 4 : 1 : 6 | **21** $1 : 6\frac{2}{3}$ |
| **22** $1 : 4\frac{1}{6}$ | **23** 1 : 2 | **24** 1 : 1.25 |
| **25** 1 : 2.5 | **26** $1 : \frac{1}{3}$ | **27** 1 : 2.4 |
| **28** 1 : 0.6 | **29** $1 : \frac{2}{3}$ | **30** $1 : \frac{7}{13}$ |

# 12   USING RATIOS

## Exercise 12A

| | | | |
|---|---|---|---|
| **1** 1 | **2** 21 | **3** 3 | **4** 20 |
| **5** 7 | **6** 15 | **7** 5 | **8** 20 |
| **9** 2 | **10** 27 | **11** 20 km/h | **12** 12 |
| **13** 20 cm | **14** 7 | **15** 98 cm | **16** 189 |
| **17** 24 | **18** $3\frac{1}{3}$ kg | **19** 22 | **20** 50 |

## Exercise 12B

| | | | |
|---|---|---|---|
| **1** 80 | **2** 8 | **3** 4 | **4** 2 |
| **5** 4 | **6** 8 | **7** 8 | **8** 2 |
| **9** 45 | **10** 30 | **11** 100 408 | |
| **12** 12.5 cm | **13** 242 | **14** 40 kg | **15** 75 |
| **16** 264 | **17** 144 seconds | | **18** 60 |
| **19** 28 | **20** 405 | | |

# 13   DIVISION USING A RATIO

## Exercise 13A

| | |
|---|---|
| **1** 36, 18 | **2** 150, 180 |
| **3** 50, 30 | **4** 42, 28 |
| **5** 128, 96 | **6** 60, 65 |
| **7** 24p, 56p | **8** 85, 102 |
| **9** 13.68 m, 17.1 m | **10** £1.00, £1.25 |
| **11** £9.60, £16.00 | **12** £4.96, £7.44 |
| **13** £6.30, £10.50 | **14** 14.25 kg, 23.75 kg |
| **15** 7.8 l, 39 l | **16** £3.24, £4.32 |
| **17** 36 kg, 18 kg | |
| **18** £14.21, £56.84, £99.47 | |
| **19** 8p, 32p, 40p | |
| **20** £5.28, £15.84, £21.12 | |
| **21** 314, $78\frac{1}{2}$, $78\frac{1}{2}$ | |
| **22** 3.45 kg, 13.8 kg, 13.8 kg | |
| **23** 150 m, 225 m, 225 m | |
| **24** 5.52 km, 16.56 km, 16.56 km, 22.08 km | |
| **25** £11.38, £28.45, £39.83, £51.21 | |

## Exercise 13B

| | |
|---|---|
| **1** 60, 40 | **2** 21, 14 |
| **3** 170, 204 | **4** 25, 15 |
| **5** £2.00, £2.50 | **6** 0.6 m, 2.4 m |
| **7** 2.4 ml, 9.6 ml | **8** 7.5 m, 1.5 m |
| **9** 36 kg, 18 kg | **10** £11.00, £13.75 |
| **11** £4.26, £10.65 | **12** £4.80, £8.00 |
| **13** £12.96, £17.28 | **14** 20.96 m, 26.2 m |
| **15** 7.8 g, 39.0 g | **16** £53.20, £85.12 |
| **17** 27, 24, 21 | |
| **18** £3.00, £4.50, £4.50 | |
| **19** £5.26, £10.52, £15.78 | |
| **20** £1, £1.50, £2 | |
| **21** 6.7 kg, 13.4 kg, 46.9 kg | |
| **22** 48 kg, 96 kg, 192 kg | |
| **23** 5.13 m, 10.26 m, 30.78 m | |
| **24** 3.52 km, 7.04 km, 14.08 km, 17.6 km | |
| **25** £2.74, £4.11, £6.85, £9.59 | |

# 14   DIRECT PROPORTION

## Exercise 14A

| | | |
|---|---|---|
| **1** £1.40 | **2** £4.50 | **3** 3.375 kg |
| **4** £1.28 | **5** 6400 km | **6** £70 |
| **7** £42.50 | **8** £112.50 | **9** 36 cm$^2$ |
| **10** £32 | **11** £3.71 | **12** 27 min |
| **13** 16 weeks | **14** 196 books | **15** 3 h |

**16** 725    **17** 2.4 cm    **18** 173.6 $l$
**19** $2\frac{1}{2}$ h    **20** £10

## Exercise 14B

**1** 120 km    **2** 288 $l$    **3** 42 s
**4** £144    **5** £310    **6** £11.25
**7** 11.9 cm    **8** £462    **9** 40p
**10** £336.60    **11** 64 km    **12** £37.20
**13** 5 boxes    **14** 72p
**15** 2800 words    **16** 35 m    **17** 20 kg
**18** 3060 km    **19** 22 units    **20** 4500 g

# 15 INVERSE PROPORTION

## Exercise 15A

**1** $13\frac{1}{2}$ days    **2** 18 weeks    **3** 25 m
**4** 10    **5** 27    **6** 4
**7** 3 h    **8** 4 days    **9** 5 h
**10** $5\frac{1}{2}$ h    **11** 12 days    **12** 3 h
**13** 27 days    **14** 30 boxes    **15** 203
**16** 6    **17** 8 days    **18** $7\frac{1}{2}$ h
**19** 96    **20** 12 days

## Exercise 15B

**1** $7\frac{1}{2}$ days    **2** $2\frac{1}{2}$ days    **3** 3 h
**4** 16 cm    **5** 6    **6** 16 days
**7** 1.5 kg    **8** 6 weeks    **9** 4 h
**10** 12 days    **11** 20 men    **12** 4 days
**13** 14    **14** 28 days    **15** 6 days
**16** 12 h    **17** 5 h    **18** 20 weeks
**19** 10 days    **20** 4

# R EVISION

## Exercise A

**1** (a) 620    (b) 0.165    (c) 14 500
   (d) 3.54    (e) 228 000    (f) 41.4
**2** (a) $\frac{5}{10}$ or 0.5    (b) $\frac{9}{1000}$ or 0.009
   (c) $\frac{4}{100}$ or 0.04    (d) 30 or $3 \times 10$
**3** (a) 2.01, 2.011, 2.1, 2.101, 2.11, 2.111
   (b) 3.007, 3.78, 3.787, 3.8, 3.9
   (c) 5.009, 5.01, 5.09, 5.1, 5.11, 5.9
**4** (a) 243    (b) 1    (c) 125
   (d) 100 000    (e) 144    (f) 141
   (g) 2
**5** (a) $2 \times 2 \times 2 \times 2 \times 3$    (b) $2 \times 2 \times 3 \times 3 \times 3$
   (c) $2 \times 3 \times 5 \times 13$
**6** (a) 1.4    (b) 17.355    (c) 6.667
   (d) 3.20

**7** (a) 112.857    (b) 4.457    (c) 8.059
**8** (a) 5.5    (b) 333    (c) 110
   (d) 50 000
**9** (a) 20    (b) 100    (c) 16
**10** (a) 2:3    (b) 1:5    (c) 4:3
   (d) 1:20    (e) 3:2    (f) 1:9
   (g) 1:6    (h) 1:3
**11** (a) 1:0.5    (b) 1:2.5    (c) 1:0.75
   (d) 1:0.5
**12** (a) 15    (b) 54
**13** (a) £4.96, £7.44    (b) 3.9 m, 19.5 m
   (c) 6.7 kg, 13.4 kg, 46.9 kg

## Exercise AA

**1** (a) (i) 22.5 m   (ii) 23.5 m
   (b) (i) 5.745 g   (ii) 5.755 g
   (c) (i) 1.5415 $l$   (ii) 1.5425 $l$
   (d) (i) 0.0285 km   (ii) 0.0295 km
**2** 5:1    **3** 2:5:8    **4** 100
**5** 36    **6** 36    **7** 28
**8** £14.70    **9** 30.4 kg    **10** 3.2 $l$
**11** 12 days    **12** 6 h    **13** 75

# 16 CANCELLING FRACTIONS

## Exercise 16A

**1** $\frac{3}{4}$    **2** $\frac{2}{3}$    **3** $\frac{2}{3}$    **4** $\frac{5}{8}$    **5** $\frac{2}{7}$
**6** $\frac{8}{9}$    **7** $\frac{4}{9}$    **8** $\frac{5}{6}$    **9** $\frac{5}{6}$    **10** $\frac{3}{5}$
**11** $\frac{3}{5}$    **12** $\frac{3}{8}$    **13** $\frac{2}{3}$    **14** $\frac{15}{16}$    **15** $\frac{2}{3}$
**16** $\frac{5}{6}$    **17** $\frac{5}{8}$    **18** $\frac{4}{7}$    **19** $\frac{40}{63}$    **20** $\frac{3}{4}$
**21** $\frac{9}{10}$    **22** $\frac{3}{4}$    **23** $\frac{3}{7}$    **24** $\frac{2}{3}$    **25** $\frac{3}{7}$
**26** $\frac{4}{7}$    **27** $\frac{5}{13}$    **28** $\frac{9}{11}$    **29** $\frac{5}{9}$    **30** $\frac{17}{20}$

## Exercise 16B

**1** $\frac{7}{12}$    **2** $\frac{1}{3}$    **3** $\frac{3}{4}$    **4** $\frac{19}{23}$    **5** $\frac{1}{3}$
**6** $\frac{3}{4}$    **7** $\frac{2}{3}$    **8** $\frac{5}{7}$    **9** $\frac{13}{17}$    **10** $\frac{3}{5}$
**11** $\frac{3}{4}$    **12** $\frac{3}{4}$    **13** $\frac{6}{7}$    **14** $\frac{5}{6}$    **15** $\frac{4}{9}$
**16** $\frac{4}{5}$    **17** $\frac{4}{5}$    **18** $\frac{7}{10}$    **19** $\frac{1}{3}$    **20** $\frac{3}{4}$
**21** $\frac{8}{9}$    **22** $\frac{3}{4}$    **23** $\frac{4}{5}$    **24** $\frac{7}{11}$    **25** $\frac{2}{3}$
**26** $\frac{2}{5}$    **27** $\frac{9}{104}$    **28** $\frac{7}{20}$    **29** $\frac{2}{5}$    **30** $\frac{11}{13}$

# 17 CONVERTING TO AND FROM MIXED NUMBERS

## Exercise 17A

1. $\frac{5}{2}$   2. $\frac{14}{3}$   3. $\frac{59}{7}$   4. $\frac{34}{5}$   5. $\frac{9}{2}$

6. $\frac{8}{3}$   7. $\frac{37}{5}$   8. $\frac{85}{11}$   9. $\frac{8}{3}$   10. $\frac{37}{10}$

11. $\frac{39}{5}$   12. $\frac{227}{12}$   13. $\frac{19}{4}$   14. $\frac{125}{8}$   15. $\frac{28}{5}$

16. $\frac{25}{7}$   17. $\frac{31}{9}$   18. $\frac{43}{5}$   19. $\frac{87}{8}$   20. $\frac{150}{11}$

21. $\frac{193}{15}$   22. $\frac{109}{9}$   23. $\frac{131}{20}$   24. $\frac{239}{15}$   25. $\frac{101}{12}$

26. $\frac{56}{15}$   27. $\frac{113}{15}$   28. $\frac{67}{14}$   29. $\frac{89}{15}$   30. $\frac{37}{16}$

## Exercise 17B

1. $\frac{15}{2}$   2. $\frac{37}{4}$   3. $\frac{19}{4}$   4. $\frac{10}{3}$   5. $\frac{26}{3}$

6. $\frac{17}{6}$   7. $\frac{33}{5}$   8. $\frac{107}{6}$   9. $\frac{86}{11}$   10. $\frac{29}{3}$

11. $\frac{53}{6}$   12. $\frac{24}{5}$   13. $\frac{14}{9}$   14. $\frac{73}{7}$   15. $\frac{77}{8}$

16. $\frac{17}{7}$   17. $\frac{160}{13}$   18. $\frac{39}{5}$   19. $\frac{128}{8}$   20. $\frac{101}{20}$

21. $\frac{181}{21}$   22. $\frac{123}{20}$   23. $\frac{194}{25}$   24. $\frac{167}{20}$   25. $\frac{191}{20}$

26. $\frac{139}{12}$   27. $\frac{277}{15}$   28. $\frac{123}{19}$   29. $\frac{85}{9}$   30. $\frac{40}{11}$

## Exercise 17C

1. $1\frac{3}{4}$   2. $1\frac{1}{3}$   3. $8\frac{1}{2}$   4. $1\frac{5}{6}$   5. $4\frac{1}{2}$

6. $2\frac{4}{5}$   7. $2\frac{1}{4}$   8. $6\frac{1}{8}$   9. $5\frac{4}{7}$   10. $17\frac{4}{5}$

11. $4\frac{9}{10}$   12. $4\frac{1}{7}$   13. $5\frac{1}{4}$   14. $4\frac{3}{5}$   15. $11\frac{3}{8}$

16. $3\frac{7}{10}$   17. $9\frac{1}{2}$   18. $4\frac{4}{5}$   19. $3\frac{3}{8}$   20. $2\frac{2}{9}$

21. $31\frac{2}{3}$   22. $3\frac{3}{5}$   23. $9\frac{2}{5}$   24. $3\frac{8}{25}$   25. $11\frac{2}{9}$

26. $19\frac{1}{5}$   27. $6\frac{13}{14}$   28. $10\frac{7}{10}$   29. $2\frac{1}{5}$   30. $8\frac{17}{25}$

## Exercise 17D

1. $3\frac{1}{5}$   2. $2\frac{1}{8}$   3. $6\frac{1}{2}$   4. $3\frac{2}{5}$   5. $2\frac{1}{3}$

6. $2\frac{5}{7}$   7. $4\frac{3}{5}$   8. $1\frac{8}{19}$   9. $3\frac{1}{2}$   10. $2\frac{2}{3}$

11. $12\frac{1}{4}$   12. $6\frac{1}{3}$   13. $1\frac{5}{6}$   14. $4\frac{2}{5}$   15. $9\frac{2}{3}$

16. $7\frac{5}{12}$   17. $5\frac{3}{10}$   18. $6\frac{5}{12}$   19. $15\frac{2}{3}$   20. $4\frac{3}{7}$

21. $7\frac{7}{10}$   22. $18\frac{1}{3}$   23. $8\frac{3}{5}$   24. $20\frac{4}{5}$   25. $6\frac{3}{16}$

26. $8\frac{9}{10}$   27. $13\frac{4}{11}$   28. $8\frac{13}{20}$   29. $8\frac{1}{11}$   30. $9\frac{1}{2}$

# 18 ADDITION OF SIMPLE FRACTIONS

## Exercise 18A

1. $1$   2. $1\frac{1}{5}$   3. $1\frac{4}{7}$   4. $1\frac{4}{9}$   5. $\frac{7}{8}$

6. $\frac{11}{16}$   7. $\frac{7}{8}$   8. $\frac{7}{16}$   9. $\frac{23}{32}$   10. $\frac{7}{16}$

11. $\frac{15}{32}$   12. $\frac{11}{32}$   13. $\frac{15}{32}$   14. $1\frac{13}{16}$   15. $1\frac{1}{16}$

16. $1\frac{1}{16}$   17. $1\frac{5}{32}$   18. $1\frac{1}{16}$   19. $1\frac{11}{16}$   20. $\frac{7}{8}$

21. $\frac{17}{20}$   22. $1\frac{1}{8}$   23. $1\frac{3}{4}$   24. $10\frac{2}{7}$   25. $5\frac{2}{9}$

26. $9$   27. $9\frac{5}{7}$   28. $8\frac{1}{8}$   29. $6\frac{4}{9}$   30. $3\frac{31}{32}$

## Exercise 18B

1. $1$   2. $1\frac{2}{9}$   3. $1\frac{3}{7}$   4. $1\frac{2}{9}$   5. $\frac{3}{8}$

6. $\frac{7}{16}$   7. $\frac{11}{16}$   8. $\frac{15}{16}$   9. $1\frac{3}{8}$   10. $\frac{9}{32}$

11. $\frac{13}{32}$   12. $1\frac{1}{8}$   13. $1\frac{1}{16}$   14. $1\frac{3}{16}$   15. $1\frac{5}{16}$

16. $1\frac{7}{32}$   17. $1\frac{1}{16}$   18. $1\frac{19}{32}$   19. $6$   20. $\frac{7}{10}$

21. $\frac{17}{24}$   22. $1\frac{1}{4}$   23. $7$   24. $8\frac{1}{8}$   25. $5$

26. $4\frac{1}{9}$   27. $9\frac{1}{7}$   28. $4\frac{1}{2}$   29. $6\frac{1}{16}$   30. $7\frac{5}{12}$

# 19 SUBTRACTION OF SIMPLE FRACTIONS

## Exercise 19A

1. $\frac{3}{4}$   2. $\frac{3}{16}$   3. $3\frac{1}{8}$   4. $2\frac{7}{8}$   5. $5\frac{1}{8}$

6. $\frac{31}{32}$   7. $3\frac{1}{2}$   8. $7\frac{1}{7}$   9. $\frac{3}{8}$   10. $\frac{3}{8}$

11. $\frac{3}{8}$   12. $\frac{1}{2}$   13. $\frac{1}{16}$   14. $\frac{1}{6}$   15. $\frac{13}{16}$

16. $\frac{1}{16}$   17. $\frac{9}{16}$   18. $\frac{1}{32}$   19. $\frac{1}{5}$   20. $\frac{3}{16}$

21. $6\frac{7}{16}$   22. $\frac{1}{8}$   23. $\frac{1}{6}$   24. $3\frac{1}{4}$   25. $8\frac{1}{2}$

26. $\frac{1}{3}$   27. $3\frac{1}{4}$   28. $1\frac{1}{16}$   29. $1\frac{1}{8}$   30. $2\frac{7}{16}$

## Exercise 19B

1. $\frac{7}{8}$   2. $1\frac{3}{4}$   3. $5\frac{1}{2}$   4. $\frac{5}{16}$   5. $6\frac{1}{8}$

6. $2\frac{3}{16}$   7. $1\frac{1}{3}$   8. $1\frac{3}{5}$   9. $\frac{2}{5}$   10. $\frac{1}{4}$

11. $\frac{1}{2}$   12. $\frac{1}{8}$   13. $\frac{1}{8}$   14. $\frac{1}{16}$   15. $\frac{15}{32}$

16. $\frac{1}{8}$   17. $\frac{1}{32}$   18. $\frac{1}{9}$   19. $2\frac{3}{10}$   20. $4\frac{1}{2}$

21. $3\frac{1}{6}$   22. $1\frac{2}{15}$   23. $4\frac{1}{6}$   24. $3\frac{1}{2}$   25. $4\frac{1}{9}$

26. $\frac{1}{16}$   27. $4\frac{3}{8}$   28. $3\frac{1}{16}$   29. $1\frac{3}{16}$   30. $1\frac{3}{8}$

# 20 SIMPLE MULTIPLICATION OF FRACTIONS

## Exercise 20A

| | | | | |
|---|---|---|---|---|
| 1 $\frac{4}{21}$ | 2 $\frac{4}{15}$ | 3 $\frac{2}{7}$ | 4 $\frac{2}{21}$ | 5 $\frac{25}{42}$ |
| 6 $\frac{5}{14}$ | 7 $\frac{1}{8}$ | 8 $\frac{2}{5}$ | 9 $1\frac{1}{7}$ | 10 $\frac{3}{5}$ |
| 11 14 | 12 $\frac{1}{2}$ | 13 $\frac{1}{32}$ | 14 $\frac{2}{3}$ | 15 $\frac{7}{18}$ |
| 16 $1\frac{1}{4}$ | 17 18 | 18 $\frac{7}{45}$ | 19 $\frac{5}{8}$ | 20 $\frac{1}{3}$ |
| 21 4 | 22 4 | 23 6 | 24 $\frac{7}{16}$ | 25 $1\frac{1}{3}$ |
| 26 $12\frac{3}{8}$ | 27 20 | 28 $\frac{27}{64}$ | 29 $5\frac{5}{6}$ | 30 $1\frac{13}{15}$ |

## Exercise 20B

| | | | | |
|---|---|---|---|---|
| 1 $\frac{2}{15}$ | 2 $2\frac{2}{9}$ | 3 $\frac{2}{35}$ | 4 $\frac{3}{10}$ | 5 $\frac{3}{10}$ |
| 6 $\frac{7}{36}$ | 7 $\frac{1}{4}$ | 8 $\frac{3}{10}$ | 9 $\frac{2}{3}$ | 10 $\frac{2}{5}$ |
| 11 $\frac{11}{20}$ | 12 $\frac{4}{7}$ | 13 24 | 14 $\frac{3}{4}$ | 15 $\frac{4}{15}$ |
| 16 1 | 17 $4\frac{4}{5}$ | 18 $\frac{1}{9}$ | 19 30 | 20 $\frac{9}{32}$ |
| 21 9 | 22 $2\frac{1}{5}$ | 23 $5\frac{1}{4}$ | 24 $\frac{5}{8}$ | 25 $\frac{21}{32}$ |
| 26 $17\frac{1}{2}$ | 27 $\frac{15}{32}$ | 28 $13\frac{1}{2}$ | 29 $3\frac{3}{4}$ | 30 $1\frac{7}{8}$ |

# 21 SIMPLE DIVISION OF FRACTIONS

## Exercise 21A

| | | | | |
|---|---|---|---|---|
| 1 $\frac{3}{10}$ | 2 $\frac{2}{3}$ | 3 $\frac{3}{4}$ | 4 $\frac{2}{9}$ | 5 $\frac{1}{16}$ |
| 6 $\frac{8}{15}$ | 7 $\frac{3}{14}$ | 8 $1\frac{1}{5}$ | 9 $1\frac{1}{8}$ | 10 $1\frac{1}{2}$ |
| 11 $1\frac{1}{14}$ | 12 $1\frac{2}{9}$ | 13 $1\frac{3}{7}$ | 14 $\frac{9}{10}$ | 15 2 |
| 16 4 | 17 $\frac{3}{17}$ | 18 $4\frac{2}{3}$ | 19 $\frac{9}{10}$ | 20 $2\frac{4}{7}$ |

## Exercise 21B

| | | | | |
|---|---|---|---|---|
| 1 $\frac{3}{4}$ | 2 $\frac{3}{8}$ | 3 $\frac{5}{8}$ | 4 $1\frac{5}{27}$ | 5 $\frac{4}{25}$ |
| 6 $\frac{5}{32}$ | 7 $\frac{27}{32}$ | 8 $\frac{9}{16}$ | 9 $\frac{7}{30}$ | 10 $3\frac{1}{4}$ |
| 11 $2\frac{1}{2}$ | 12 $2\frac{5}{8}$ | 13 $\frac{2}{3}$ | 14 $1\frac{5}{8}$ | 15 $\frac{1}{2}$ |
| 16 $1\frac{1}{9}$ | 17 $\frac{1}{30}$ | 18 $\frac{2}{3}$ | 19 $1\frac{1}{5}$ | 20 $1\frac{1}{5}$ |

# 22 CONVERTING BETWEEN FRACTIONS AND DECIMALS

## Exercise 22A

| | | |
|---|---|---|
| 1 0.75 | 2 0.875 | 3 0.1875 |
| 4 0.4 | 5 0.03125 | 6 0.9 |
| 7 4.08 | 8 4.81 | 9 9.6 |
| 10 0.55 | 11 3.44 | 12 2.1 |
| 13 0.6875 | 14 8.44 | 15 7.03 |
| 16 5.98 | 17 3.3125 | 18 2.12 |
| 19 3.05 | 20 1.02 | 21 0.4333 |
| 22 0.5556 | 23 0.4286 | 24 0.2222 |
| 25 0.4167 | 26 0.3333 | 27 0.5455 |
| 28 0.1818 | 29 0.4444 | 30 0.2727 |

## Exercise 22B

| | | |
|---|---|---|
| 1 0.625 | 2 0.6 | 3 0.4375 |
| 4 0.1 | 5 3.7 | 6 7.5 |
| 7 0.35 | 8 2.2 | 9 2.225 |
| 10 1.9 | 11 3.8 | 12 0.85 |
| 13 3.16 | 14 3.52 | 15 2.325 |
| 16 1.9375 | 17 0.406 25 | 18 4.0625 |
| 19 3.68 | 20 7.98 | 21 0.4545 |
| 22 0.9091 | 23 0.6333 | 24 0.6667 |
| 25 0.5714 | 26 0.5556 | 27 0.0909 |
| 28 0.8889 | 29 0.5833 | 30 0.0833 |

## Exercise 22C

| | | | | |
|---|---|---|---|---|
| 1 $\frac{13}{100}$ | 2 $\frac{1}{10}$ | 3 $\frac{3}{4}$ | 4 $\frac{9}{20}$ | 5 $\frac{1}{5}$ |
| 6 $\frac{9}{100}$ | 7 $\frac{5}{8}$ | 8 $\frac{377}{1000}$ | 9 $\frac{3}{50}$ | 10 $\frac{2}{25}$ |
| 11 $\frac{11}{25}$ | 12 $\frac{1}{1000}$ | 13 $\frac{11}{20}$ | 14 $\frac{93}{100}$ | 15 $\frac{3}{500}$ |
| 16 $\frac{27}{40}$ | 17 $\frac{22}{25}$ | 18 $\frac{21}{25}$ | 19 $1\frac{9}{25}$ | 20 $\frac{179}{200}$ |
| 21 $3\frac{13}{16}$ | 22 $7\frac{2}{5}$ | 23 $\frac{1}{40}$ | 24 $\frac{19}{25}$ | 25 $5\frac{1}{20}$ |
| 26 $\frac{151}{200}$ | 27 $8\frac{7}{8}$ | 28 $\frac{6}{25}$ | 29 $4\frac{13}{25}$ | 30 $3\frac{49}{80}$ |

## Exercise 22D

| | | | | |
|---|---|---|---|---|
| 1 $\frac{3}{10}$ | 2 $\frac{57}{100}$ | 3 $\frac{3}{20}$ | 4 $\frac{8}{25}$ | 5 $\frac{9}{10}$ |
| 6 $\frac{17}{40}$ | 7 $\frac{999}{1000}$ | 8 $9\frac{3}{10}$ | 9 $\frac{4}{25}$ | 10 $\frac{47}{100}$ |
| 11 $\frac{103}{1000}$ | 12 $\frac{33}{40}$ | 13 $\frac{11}{40}$ | 14 $\frac{13}{25}$ | 15 $\frac{87}{200}$ |
| 16 $7\frac{4}{5}$ | 17 $\frac{2}{25}$ | 18 $5\frac{1}{25}$ | 19 $\frac{9}{25}$ | 20 $\frac{7}{8}$ |
| 21 $2\frac{17}{80}$ | 22 $\frac{163}{200}$ | 23 $2\frac{7}{40}$ | 24 $4\frac{15}{16}$ | 25 $\frac{7}{25}$ |
| 26 $\frac{143}{200}$ | 27 $\frac{171}{200}$ | 28 $2\frac{3}{32}$ | 29 $\frac{13}{40}$ | 30 $2\frac{13}{16}$ |

# 23 CONVERTING BETWEEN DECIMALS AND PERCENTAGES

## Exercise 23A

| | | | |
|---|---|---|---|
| 1 20% | 2 $12\frac{1}{2}$% | 3 23% | 4 38% |
| 5 40% | 6 $14\frac{1}{2}$% | 7 93% | 8 80% |
| 9 $25\frac{1}{2}$% | 10 46% | 11 13% | 12 78% |
| 13 $21\frac{1}{2}$% | 14 48% | 15 $66\frac{1}{2}$% | 16 $16\frac{1}{2}$% |
| 17 86% | 18 $4\frac{3}{4}$% | 19 $17\frac{1}{4}$% | 20 64% |
| 21 65% | 22 $12\frac{1}{2}$% | 23 $45\frac{1}{2}$% | 24 $2\frac{1}{4}$% |
| 25 $\frac{1}{4}$% | 26 $1\frac{1}{2}$% | 27 $65\frac{1}{4}$% | 28 $92\frac{3}{4}$% |
| 29 $4\frac{1}{2}$% | 30 $75\frac{3}{4}$% | | |

## Exercise 23B

| | | | | | | | |
|---|---|---|---|---|---|---|---|
| **1** | $42\frac{1}{2}$% | **2** | 50% | **3** | $35\frac{1}{2}$% | **4** | 17% |
| **5** | 53% | **6** | 60% | **7** | 98% | **8** | 39% |
| **9** | 90% | **10** | 66% | **11** | $35\frac{1}{2}$% | **12** | 21% |
| **13** | 56% | **14** | $83\frac{1}{2}$% | **15** | $29\frac{1}{4}$% | **16** | $77\frac{3}{4}$% |
| **17** | 61% | **18** | $65\frac{1}{2}$% | **19** | 83% | **20** | 95% |
| **21** | 45% | **22** | $83\frac{1}{2}$% | **23** | 9% | **24** | $82\frac{1}{2}$% |
| **25** | $3\frac{1}{2}$% | **26** | $\frac{3}{4}$% | **27** | $73\frac{1}{2}$% | **28** | $4\frac{1}{4}$% |
| **29** | $19\frac{3}{4}$% | **30** | $55\frac{1}{4}$% | | | | |

## Exercise 23C

| | | | | | | | |
|---|---|---|---|---|---|---|---|
| **1** | 0.31 | **2** | 0.85 | **3** | 0.1 | **4** | 0.42 |
| **5** | 0.16 | **6** | 0.5 | **7** | 0.57 | **8** | 0.02 |
| **9** | 0.96 | **10** | 0.41 | **11** | 0.18 | **12** | 0.2175 |
| **13** | 0.865 | **14** | 0.44 | **15** | 0.1525 | **16** | 0.62 |
| **17** | 0.3225 | **18** | 0.79 | **19** | 0.3175 | **20** | 0.495 |
| **21** | 0.335 | **22** | 0.18 | **23** | 0.4875 | **24** | 0.685 |
| **25** | 0.945 | **26** | 0.127 | **27** | 0.8542 | **28** | 0.221 |
| **29** | 0.199 | **30** | 0.4435 | | | | |

## Exercise 23D

| | | | | | | | |
|---|---|---|---|---|---|---|---|
| **1** | 0.86 | **2** | 0.6 | **3** | 0.31 | **4** | 0.03 |
| **5** | 0.53 | **6** | 0.3 | **7** | 0.82 | **8** | 0.42 |
| **9** | 0.08 | **10** | 0.9 | **11** | 0.5425 | **12** | 0.04 |
| **13** | 0.175 | **14** | 0.2975 | **15** | 0.69 | **16** | 0.59 |
| **17** | 0.5175 | **18** | 0.72 | **19** | 0.8025 | **20** | 0.595 |
| **21** | 0.135 | **22** | 0.4225 | **23** | 0.095 | **24** | 0.0075 |
| **25** | 0.145 | **26** | 0.737 | **27** | 0.848 | **28** | 0.1435 |
| **29** | 0.9255 | **30** | 0.1494 | | | | |

# 24 CONVERTING BETWEEN FRACTIONS AND PERCENTAGES

## Exercise 24A

| | | | | | | | |
|---|---|---|---|---|---|---|---|
| **1** | 40% | **2** | 45% | **3** | 28% | **4** | $2\frac{1}{2}$% |
| **5** | 72% | **6** | 17% | **7** | 16% | **8** | 65% |
| **9** | 62% | **10** | $27\frac{1}{2}$% | **11** | $\frac{1}{2}$% | **12** | 95% |
| **13** | $33\frac{1}{3}$% | **14** | 15% | **15** | 68% | **16** | $12\frac{1}{2}$% |
| **17** | $\frac{3}{4}$% | **18** | 16% | **19** | $22\frac{2}{9}$% | **20** | $32\frac{1}{2}$% |
| **21** | 18% | **22** | 60% | **23** | 44% | **24** | 76% |
| **25** | $26\frac{1}{4}$% | **26** | $73\frac{3}{4}$% | **27** | $62\frac{1}{2}$% | **28** | $88\frac{8}{9}$% |
| **29** | $58\frac{1}{3}$% | **30** | $91\frac{2}{3}$% | | | | |

## Exercise 24B

| | | | | | | | |
|---|---|---|---|---|---|---|---|
| **1** | 80% | **2** | 50% | **3** | 64% | **4** | 55% |
| **5** | 11% | **6** | 56% | **7** | 85% | **8** | 84% |
| **9** | 35% | **10** | $7\frac{1}{2}$% | **11** | $57\frac{1}{2}$% | **12** | $8\frac{1}{2}$% |
| **13** | $66\frac{2}{3}$% | **14** | $17\frac{1}{2}$% | **15** | 5% | **16** | 52% |
| **17** | 48% | **18** | $16\frac{2}{3}$% | **19** | $3\frac{3}{4}$% | **20** | $44\frac{4}{9}$% |
| **21** | 12% | **22** | $22\frac{1}{2}$% | **23** | $8\frac{3}{4}$% | **24** | $78\frac{3}{4}$% |
| **25** | $97\frac{1}{2}$% | **26** | $83\frac{1}{3}$% | **27** | $77\frac{7}{9}$% | **28** | $87\frac{1}{2}$% |
| **29** | $8\frac{1}{3}$% | **30** | $41\frac{2}{3}$% | | | | |

## Exercise 24C

| | | | | | | | | | |
|---|---|---|---|---|---|---|---|---|---|
| **1** | $\frac{1}{5}$ | **2** | $\frac{7}{100}$ | **3** | $\frac{1}{2}$ | **4** | $\frac{7}{10}$ | **5** | $\frac{29}{100}$ |
| **6** | $\frac{19}{20}$ | **7** | $\frac{57}{100}$ | **8** | $\frac{9}{10}$ | **9** | $\frac{11}{100}$ | **10** | $\frac{9}{25}$ |
| **11** | $\frac{3}{20}$ | **12** | $\frac{18}{25}$ | **13** | $\frac{9}{20}$ | **14** | $\frac{4}{25}$ | **15** | $\frac{52}{100}$ |
| **16** | $\frac{1}{300}$ | **17** | $\frac{19}{40}$ | **18** | $\frac{1}{80}$ | **19** | $\frac{13}{40}$ | **20** | $\frac{2}{3}$ |
| **21** | $\frac{27}{40}$ | **22** | $\frac{191}{200}$ | **23** | $\frac{37}{80}$ | **24** | $\frac{359}{400}$ | **25** | $\frac{47}{60}$ |
| **26** | $\frac{26}{75}$ | **27** | $\frac{113}{800}$ | **28** | $\frac{97}{150}$ | **29** | $\frac{87}{400}$ | **30** | $\frac{97}{300}$ |

## Exercise 24D

| | | | | | | | | | |
|---|---|---|---|---|---|---|---|---|---|
| **1** | $\frac{2}{5}$ | **2** | $\frac{9}{100}$ | **3** | $\frac{37}{100}$ | **4** | $\frac{9}{20}$ | **5** | $\frac{51}{100}$ |
| **6** | $\frac{3}{25}$ | **7** | $\frac{17}{20}$ | **8** | $\frac{19}{100}$ | **9** | $\frac{6}{25}$ | **10** | $\frac{1}{20}$ |
| **11** | $\frac{8}{25}$ | **12** | $\frac{7}{20}$ | **13** | $\frac{12}{25}$ | **14** | $\frac{19}{20}$ | **15** | $\frac{13}{20}$ |
| **16** | $\frac{1}{7}$ | **17** | $\frac{5}{12}$ | **18** | $\frac{3}{400}$ | **19** | $\frac{31}{500}$ | **20** | $\frac{1}{3}$ |
| **21** | $\frac{79}{200}$ | **22** | $\frac{11}{200}$ | **23** | $\frac{167}{400}$ | **24** | $\frac{349}{400}$ | **25** | $\frac{23}{160}$ |
| **26** | $\frac{61}{75}$ | **27** | $\frac{1}{15}$ | **28** | $\frac{293}{800}$ | **29** | $\frac{81}{200}$ | **30** | $\frac{37}{150}$ |

# 25 FRACTIONAL CHANGES

## Exercise 25A

| | | | | | | | |
|---|---|---|---|---|---|---|---|
| **1** | 20 | **2** | 40 | **3** | 9 | **4** | 63 |
| **5** | 15 | **6** | 6 | **7** | 35 l | **8** | 36 |
| **9** | 10 | **10** | 50 | **11** | 16 m | **12** | 80 |
| **13** | 50 | **14** | 12 | **15** | 39 g | **16** | 56 m |
| **17** | 60 km | **18** | 99 | **19** | 28 km | **20** | £2.50 |
| **21** | 143 kg | **22** | 10 m | **23** | 25 kg | **24** | 70p |
| **25** | $70 | **26** | 165 | **27** | 85p | **28** | 68 |
| **29** | £55.25 | **30** | £4.56 | | | | |

## Exercise 25B

| | | | | | | | |
|---|---|---|---|---|---|---|---|
| **1** | 15 | **2** | 36 | **3** | 18 | **4** | 16 |
| **5** | 4 | **6** | 4 | **7** | 12 | **8** | 12 |
| **9** | 26 | **10** | 5 | **11** | 10 | **12** | 91 |
| **13** | 51 | **14** | 70 | **15** | 18 | **16** | $14 |
| **17** | 55 l | **18** | $25 | **19** | 5 m | **20** | £165 |
| **21** | 30p | **22** | £28 | **23** | 32 t | **24** | 4 g |
| **25** | 85 m | **26** | 49 cm | **27** | 220 g | **28** | 20 cm |
| **29** | £95 | **30** | £15.06 | | | | |

# 26 ONE NUMBER AS A FRACTION OF ANOTHER

## Exercise 26A

**1** $\frac{1}{13}$    **2** $\frac{9}{25}$    **3** $\frac{1}{8}$    **4** $\frac{3}{13}$    **5** $\frac{1}{8}$

**6** $\frac{5}{7}$    **7** $\frac{1}{2}$    **8** $\frac{7}{17}$    **9** $\frac{3}{13}$    **10** $\frac{2}{7}$

**11** $\frac{1}{5}$    **12** $\frac{25}{27}$    **13** $\frac{8}{15}$    **14** $\frac{11}{24}$    **15** $\frac{3}{11}$

**16** $\frac{3}{7}$    **17** $\frac{8}{25}$    **18** $\frac{24}{65}$    **19** $\frac{1}{3}$    **20** $\frac{2}{5}$

## Exercise 26B

**1** $\frac{1}{3}$    **2** $\frac{3}{17}$    **3** $\frac{1}{6}$    **4** $\frac{1}{2}$    **5** $\frac{2}{7}$

**6** $\frac{3}{7}$    **7** $\frac{2}{9}$    **8** $\frac{1}{6}$    **9** $\frac{1}{4}$    **10** $\frac{6}{13}$

**11** $\frac{1}{2}$    **12** $\frac{4}{15}$    **13** $\frac{10}{111}$    **14** $\frac{3}{7}$    **15** $\frac{1}{3}$

**16** $\frac{1}{9}$    **17** $\frac{3}{13}$    **18** $\frac{2}{5}$    **19** $\frac{2}{3}$    **20** $\frac{11}{30}$

# 27 PERCENTAGE CHANGE

## Exercise 27A

**1** 800    **2** £10.50    **3** £34
**4** 285    **5** 1326    **6** £59.70
**7** £436.80    **8** 44 km    **9** 5.85 km
**10** 708 g    **11** £2.93    **12** 154 ml
**13** 720 g    **14** 1287 km    **15** 237.5 g
**16** £43.68    **17** 102 ml    **18** £2.43
**19** £1.65    **20** 1256 m    **21** £10.28
**22** £19.38    **23** £240    **24** £7.03
**25** £6.17    **26** £94.71    **27** £31.50
**28** £291.60    **29** £140    **30** £7494.25

## Exercise 27B

**1** 580    **2** 280    **3** 80p
**4** £249.60    **5** £228    **6** £37.93
**7** £2.10    **8** £39.95    **9** 2190 ml
**10** £736    **11** £28.64    **12** 48 km
**13** 384 cm    **14** 1540 mm    **15** 57 g
**16** 1360 kg    **17** £1.13    **18** £6.00
**19** £16.43    **20** £16.88    **21** £1.60
**22** £91.38    **23** 49p    **24** £2.84
**25** £2.31    **26** £152.74    **27** £28.34
**28** £147.82    **29** £896.75    **30** 5795

# 28 ONE NUMBER AS A PERCENTAGE OF ANOTHER

## Exercise 28A

**1** 5%    **2** 40%    **3** $37\frac{1}{2}$%    **4** 60%
**5** 20%    **6** 16%    **7** 32%    **8** 30%

**9** 32%    **10** 20%    **11** 4%    **12** 14%
**13** 40%    **14** 4%    **15** 30%    **16** 25%
**17** $6\frac{1}{4}$%    **18** $12\frac{1}{2}$%    **19** $3\frac{1}{3}$%    **20** $27\frac{1}{5}$%
**21** $16\frac{2}{3}$%    **22** 30%    **23** $62\frac{1}{2}$%    **24** $37\frac{1}{2}$%
**25** 20%    **26** 24%    **27** 25%    **28** $93\frac{1}{3}$%
**29** 95%    **30** $13\frac{1}{3}$%

## Exercise 28B

**1** 1%    **2** 75%    **3** $67\frac{1}{2}$%    **4** 40%
**5** 65%    **6** 15%    **7** 8%    **8** 28%
**9** 20%    **10** 45%    **11** 60%    **12** $10\frac{4}{5}$%
**13** 95%    **14** $6\frac{2}{3}$%    **15** 70%    **16** 35%
**17** $6\frac{1}{4}$%    **18** $17\frac{1}{2}$%    **19** $41\frac{2}{3}$%    **20** $22\frac{1}{2}$%
**21** $17\frac{1}{2}$%    **22** 37%    **23** $6\frac{2}{3}$%    **24** $86\frac{2}{3}$%
**25** $12\frac{1}{2}$%    **26** 40%    **27** 90%    **28** 20%
**29** 54%    **30** $43\frac{3}{4}$%

# REVISION

## Exercise B

**1** (a) $\frac{3}{4}$   (b) $\frac{5}{11}$   (c) $\frac{4}{5}$   (d) $\frac{3}{8}$   (e) $\frac{2}{5}$

**2** (a) $\frac{31}{4}$   (b) $\frac{34}{7}$   (c) $\frac{28}{5}$   (d) $\frac{85}{9}$

**3** (a) $7\frac{3}{10}$   (b) $7\frac{2}{3}$   (c) $2\frac{1}{5}$   (d) $6\frac{6}{7}$

**4** (a) $1\frac{1}{2}$   (b) $\frac{21}{32}$   (c) $9\frac{4}{9}$   (d) $\frac{4}{7}$   (e) $1\frac{9}{16}$

**5** (a) $\frac{7}{20}$   (b) 8   (c) $5\frac{5}{8}$   (d) $1\frac{3}{4}$   (e) $\frac{1}{18}$

**6** (a) 0.15   (b) 2.36   (c) 0.37   (d) 0.175

**7** (a) $\frac{3}{8}$   (b) $1\frac{5}{16}$   (c) $\frac{14}{25}$   (d) $\frac{3}{8}$

**8** (a) 39%   (b) $81\frac{1}{2}$%   (c) 42%   (d) $87\frac{1}{2}$%

**9** (a) 204 g   (b) £14.70

**10** (a) 30   (b) 1179 km

**11** (a) 20   (b) £2.31

## Exercise BB

**1** £101.05    **2** £26.38    **3** £361.25
**4** £10 658.36    **5** £176.25    **6** 2484
**7** $12\frac{1}{2}$%    **8** 25%    **9** $33\frac{1}{3}$%
**10** $93\frac{1}{2}$%    **11** £360    **12** £108.40
**13** (a) £1089.90   (b) £143.90
**14** (a) £308.08   (b) £54.08
**15** (a) £3989.80   (b) £598.80   (c) 18.15%

# Algebra

## 29 CONTINUING A NUMBER SEQUENCE

### Exercise 29A

| | | |
|---|---|---|
| **1** 34, 42 | **2** 58, 69 | **3** 18, 12 |
| **4** 14, 5 | **5** 45, 51 | **6** 28, 37 |
| **7** 21, 29 | **8** 32, 41 | **9** 55, 75 |
| **10** 12, 7 | **11** 9, 0 | **12** 49, 64 |
| **13** 23, 33 | **14** 24, 26 | **15** 45, 58 |
| **16** 12, 2 | **17** 46, 68 | **18** 55, 61 |
| **19** 22, 34 | **20** 8, 10 | **21** 38, 43 |
| **22** 65, 84 | **23** 15, 3 | **24** 48, 69 |
| **25** 2, 4 | **26** 64, 100 | **27** 15, 14 |
| **28** 21, 32 | **29** 102, 129 | **30** 64, 92 |

### Exercise 29B

| | | |
|---|---|---|
| **1** 31, 38 | **2** 50, 59 | **3** 9, 4 |
| **4** 12, 4 | **5** 34, 43 | **6** 32, 42 |
| **7** 56, 81 | **8** 59, 67 | **9** 22, 34 |
| **10** 6, 0 | **11** 76, 140 | **12** 17, 6 |
| **13** 29, 32 | **14** 31, 43 | **15** 26, 29 |
| **16** 11, 3 | **17** 59, 83 | **18** 11, 13 |
| **19** 4, 7 | **20** 50, 73 | **21** 30, 32 |
| **22** 30, 41 | **23** 70, 78 | **24** 24, 23 |
| **25** 69, 72 | **26** 49, 62 | **27** 13, 8 |
| **28** 43, 64 | **29** 40, 42 | **30** 34, 46 |

## 30 MAKING PREDICTIONS AND GENERALISING IN NUMBER SERIES

### Exercise 30A

In the following the rule has been given in symbols, but could easily be translated into words for part (a).

| | | |
|---|---|---|
| **1** (b) $3n - 2$ | (c) 28, 43 |
| **2** (b) $2n - 3$ | (c) 17, 27 |
| **3** (b) $3n - 5$ | (c) 25, 40 |
| **4** (b) $5n + 3$ | (c) 53, 78 |
| **5** (b) $4n + 5$ | (c) 65, 85 |
| **6** (b) $3n + 3$ | (c) 48, 63 |
| **7** (b) $2n - 5$ | (c) 25, 35 |
| **8** (b) $4n - 1$ | (c) 39, 79 |
| **9** (b) $3n + 4$ | (c) 34, 64 |
| **10** (b) $5n - 4$ | (c) 71, 96 |
| **11** (b) $4n + 3$ | (c) 63, 83 |
| **12** (b) $5n - 2$ | (c) 73, 98 |
| **13** (b) $4n - 4$ | (c) 36, 76 |
| **14** (b) $6n - 1$ | (c) 59, 119 |
| **15** (b) $5n + 1$ | (c) 51, 76 |
| **16** (b) $6n + 2$ | (c) 62, 92 |
| **17** (b) $30 - n$ | (c) 15, 10 |
| **18** (b) $3n - 6$ | (c) 24, 39 |
| **19** (b) $5n + 4$ | (c) 54, 79 |
| **20** (b) $50 - 2n$ | (c) 20, 10 |
| **21** (b) $n^2 + 4$ | (c) 104, 404 |
| **22** (b) $n^2 - 1$ | (c) 99, 399 |
| **23** (b) $n^2 + 3$ | (c) 103, 403 |
| **24** (b) $2n^2 + 3n$ | (c) 230, 860 |
| **25** (b) $n^2 + 2n$ | (c) 120, 440 |
| **26** (b) $2n^2 + 1$ | (c) 201, 801 |
| **27** (b) $3n^2 + n$ | (c) 690, 1220 |
| **28** (b) $n^2 - n + 4$ | (c) 214, 384 |
| **29** (b) $n^2 + 2n - 1$ | (c) 254, 43 |
| **30** (b) $n^2 + 3n + 2$ | (c) 272, 462 |

### Exercise 30B

In the following the rule has been given in symbols, but could easily be translated into words for part (a).

| | | |
|---|---|---|
| **1** (b) $4n + 1$ | (c) 41, 61 |
| **2** (b) $5n - 3$ | (c) 47, 72 |
| **3** (b) $2n - 2$ | (c) 18, 28 |
| **4** (b) $2n + 9$ | (c) 29, 39 |
| **5** (b) $4n + 6$ | (c) 46, 66 |
| **6** (b) $2n + 7$ | (c) 37, 47 |
| **7** (b) $4n + 4$ | (c) 64, 84 |
| **8** (b) $5n + 2$ | (c) 77, 102 |
| **9** (b) $4n + 2$ | (c) 42, 82 |
| **10** (b) $2n - 7$ | (c) 13, 33 |
| **11** (b) $3n - 1$ | (c) 29, 59 |
| **12** (b) $4n - 2$ | (c) 58, 78 |
| **13** (b) $3n + 2$ | (c) 47, 62 |
| **14** (b) $8n - 1$ | (c) 119, 159 |
| **15** (b) $3n + 6$ | (c) 36, 66 |
| **16** (b) $5n - 1$ | (c) 74, 99 |
| **17** (b) $6n + 3$ | (c) 93, 123 |
| **18** (b) $25 - n$ | (c) 15, 5 |
| **19** (b) $2n + 8$ | (c) 38, 48 |
| **20** (b) $45 - 2n$ | (c) 25, 5 |
| **21** (b) $4n - 5$ | (c) 55, 75 |
| **22** (b) $n^2 + 3$ | (c) 103, 403 |
| **23** (b) $n^2 - 2$ | (c) 98, 398 |
| **24** (b) $n^2 + 6$ | (c) 106, 406 |

**25** (b) $2n^2 - n$     (c) 190, 780
**26** (b) $n^2 + 4n$     (c) 285, 480
**27** (b) $3n^2 + 3$     (c) 678, 1203
**28** (b) $2n^2 + n$     (c) 465, 820
**29** (b) $n^2 + n - 2$     (c) 238, 418
**30** (b) $n^2 + 2n + 1$     (c) 256, 441

## 31 SERIES FROM DIAGRAMS

### Exercise 31A

| | | |
|---|---|---|
| **1** $2n + 1$ | **2** $2n + 2$ | **3** $3n + 3$ |
| **4** $2n + 2$ | **5** $5n + 1$ | **6** $4n + 4$ |
| **7** $4n + 2$ | **8** $3n + 5$ | **9** $5n + 1$ |
| **10** $2n - 1$ | **11** $3n + 2$ | **12** $n(n + 2)$ |
| **13** $n(n + 4)$ | **14** $n^2$ | **15** $\frac{1}{2}n(n + 1)$ |

### Exercise 31B

| | | |
|---|---|---|
| **1** $2n + 2$ | **2** $5n + 2$ | **3** $n - 1$ |
| **4** $n + 2$ | **5** $2n - 2$ | **6** $3n + 3$ |
| **7** $7n + 3$ | **8** $5n + 3$ | **9** $4n + 2$ |
| **10** $2n + 4$ | **11** $8n + 5$ | **12** $9n + 2$ |
| **13** $2n + 1$ | **14** $\frac{1}{2}n(n + 1)$ | **15** $2n^2 - n$ |

## 32 COLLECTING LIKE TERMS

### Exercise 32A

| | | |
|---|---|---|
| **1** $2a + 3b$ | **2** $2a + b + 4c$ | **3** $5a + 3b + c$ |
| **4** $4a$ | **5** $m + 4n$ | **6** $5a + c$ |
| **7** $2y$ | **8** $2a + 2c$ | **9** $2a + c$ |
| **10** $2e$ | **11** $2a + c$ | **12** $5ab + 2a + 3b$ |
| **13** $2cd + 4xy$ | **14** $4x^3 + x^2$ | **15** $3y^2 + 4x + y$ |
| **16** $8cd^2$ | **17** $8a^3$ | **18** $4a^3y^2 + 4a^2y^2$ |
| **19** $2a^3 + a^2 + 2a$ | | **20** $8abc + 2ab + a$ |
| **21** $xyuv$ | | **22** $5t^3 + 9t^2$ |
| **23** $3a^2$ | | **24** $3pqr$ |
| **25** $4a^5 + 3a^2 + a$ | | **26** $8a^3 - a^4 + 4a^2$ |
| **27** $2x^2 + 5x - 2$ | | **28** $8g^3 + 10g^2 - 3$ |
| **29** $6x^3 - 3x^2$ | | **30** $3a^5 + 2a^4 + 2a^3 + a^2$ |

### Exercise 32B

| | | |
|---|---|---|
| **1** $2a + 2b + c$ | **2** $5a + c$ | **3** $3b + 3c - 2a$ |
| **4** $2a + 3b - c$ | **5** $5w + 2y$ | **6** $w + 6x + 2y$ |
| **7** $2a + 2c$ | **8** $3a + b$ | **9** $4n + p$ |
| **10** $3b$ | **11** $0$ | **12** $3xy + x + 2y$ |
| **13** $2cd + d$ | | **14** $6x^2 + x$ |
| **15** $x^3$ | | **16** $5x^2 + 4y^2 + 5y$ |
| **17** $4c^2d^3 + 5c^3d^3$ | | **18** $6st^2$ |
| **19** $5xyuv$ | | **20** $4a^2$ |
| **21** $7pqr + p$ | | **22** $4a^3 + 3a^2 + 4a$ |

**23** $5w^3 + 4w^2$     **24** $pqr$
**25** $a^4 + 3a^2 + a$     **26** $x^2 - 2x^3 - 3x^5$
**27** $6x^3 - x^2 - 6$     **28** $5t^3 + t^2 + 5$
**29** $6a^3 - a^4 + 3a^2$     **30** $4x^2 + 9x - 3$

## 33 MULTIPLICATION OF TERMS

### Exercise 33A

| | | | |
|---|---|---|---|
| **1** $20mn$ | **2** $24abc$ | **3** $30abc$ | **4** $20xy$ |
| **5** $40abc$ | **6** $4a^3$ | **7** $49q^2$ | **8** $5a^2b^2$ |
| **9** $9x^2y^2$ | **10** $6c^3d^2$ | **11** $x^7$ | **12** $t^{10}$ |
| **13** $y^9$ | **14** $n^6$ | **15** $d^{12}$ | **16** $6x^8$ |
| **17** $21f^9$ | **18** $30y^{11}$ | **19** $12c^{13}$ | **20** $24x^{13}$ |
| **21** $9a^2$ | **22** $gh^2i$ | **23** $3w^2y$ | **24** $16c^2d$ |
| **25** $6hk^2m$ | **26** $4d^2e^2$ | **27** $9w^2x^2y$ | **28** $1000a^3$ |
| **29** $4y^4$ | **30** $6y^3x$ | | |

### Exercise 33B

| | | | |
|---|---|---|---|
| **1** $21qr$ | **2** $32pq$ | **3** $12acd$ | **4** $24abc$ |
| **5** $75ntw$ | **6** $y^5$ | **7** $x^3y^2$ | **8** $4a^2b$ |
| **9** $25ay^3$ | **10** $16g^3$ | **11** $a^2b^3$ | **12** $9a^2b$ |
| **13** $x^9$ | **14** $e^{19}$ | **15** $m^{16}$ | **16** $12y^5$ |
| **17** $20g^{11}$ | **18** $24d^6$ | **19** $6x^{19}$ | **20** $24e^{14}$ |
| **21** $8w^{13}$ | **22** $16d^2$ | **23** $s^2t^2$ | **24** $2a^2b$ |
| **25** $e^2fg$ | **26** $12a^2bc$ | **27** $20g^2h^2$ | **28** $6a^2b^2$ |
| **29** $6a^3b$ | **30** $a^5b^2$ | | |

## 34 DIVISION OF TERMS

### Exercise 34A

| | | | |
|---|---|---|---|
| **1** $f$ | **2** $2m$ | **3** $1$ | **4** $\frac{2r}{3}$ |
| **5** $m$ | **6** $km$ | **7** $2x$ | **8** $4$ |
| **9** $3$ | **10** $3a$ | **11** $x$ | **12** $y^8$ |
| **13** $g^{11}$ | **14** $5x^4$ | **15** $3m^7$ | **16** $4t^8$ |
| **17** $5$ | **18** $2a^5b^4$ | **19** $\frac{a}{2}$ | **20** $3p^2q$ |
| **21** $\frac{4a}{5}$ | **22** $\frac{2r}{3}$ | **23** $2a^2b^5$ | **24** $\frac{5k}{h}$ |
| **25** $\frac{8a}{3d}$ | **26** $\frac{x}{2}$ | **27** $\frac{2y}{3x}$ | **28** $\frac{3tu}{2s}$ |
| **29** $\frac{3p^3q^2}{2}$ | **30** $\frac{x}{2y}$ | | |

### Exercise 34B

| | | | |
|---|---|---|---|
| **1** $\frac{3g}{2}$ | **2** $p$ | **3** $\frac{d}{e}$ | **4** $\frac{1}{g}$ |
| **5** $n$ | **6** $2a^2$ | **7** $2p$ | **8** $12$ |

| **9** $2c$ | **10** $x^7$ | **11** $k^6$ | **12** $t^4$ |
|---|---|---|---|
| **13** $w^6$ | **14** $3y^5$ | **15** $2b^8$ | **16** $2a^2b$ |
| **17** $2x$ | **18** $4p^2$ | **19** $2ab$ | **20** $2x^2y^3$ |
| **21** $2f$ | **22** $4a^2b^2$ | **23** $\dfrac{a}{bc}$ | **24** $\dfrac{m}{2}$ |
| **25** $3cd$ | **26** $\dfrac{3p^3q^2}{2}$ | **27** $\dfrac{2m}{3}$ | **28** $\dfrac{p}{3}$ |
| **29** $\dfrac{x}{y}$ | **30** $\dfrac{a}{bc}$ | | |

## 35 BRACKETS

### Exercise 35A

| | | | |
|---|---|---|---|
| **1** $11b - 9$ | | **2** $11c + 4$ | |
| **3** $2x + 3y$ | | **4** $x - y$ | |
| **5** $d - 11$ | | **6** $9p - 3q$ | |
| **7** $2y^2 + x^2$ | | **8** $11x - 3$ | |
| **9** $a^2 + ab + b^2$ | | **10** $8x^2 - 15x + 10$ | |
| **11** $10c - 14$ | | **12** $1 - 17t$ | |
| **13** $9xy + 2wx$ | | **14** $2wx + 10wy$ | |
| **15** $7ab$ | | **16** $2a + b$ | |
| **17** $2p + q$ | | **18** $a - b$ | |
| **19** $4mp - 2np - 6np$ | | **20** $9p - 3q$ | |
| **21** $2y^2 + x^2$ | | **22** $3pm + 2pn - 2mn$ | |
| **23** $2xy + 2wx - wy$ | | **24** $a - 3$ | |
| **25** $5 - 2a$ | | **26** $2x + 8$ | |
| **27** $29 - 3x$ | | **28** $2y^2 + 7y$ | |
| **29** $23 - 8x$ | | **30** $7t - t^2$ | |

### Exercise 35B

| | | | |
|---|---|---|---|
| **1** $7r - 7t$ | | **2** $5x + 15$ | |
| **3** $4a + 11$ | | **4** $k - 2m$ | |
| **5** $x^2 + xy + y^2$ | | **6** $5x + 1$ | |
| **7** $2y - 2t$ | | **8** $x^2 + 5x - 12$ | |
| **9** $2d^2 + 3c^2$ | | **10** $2c^2 + 3d^2$ | |
| **11** $9x^2 - 4y^2$ | | **12** $a^2 + b^2 + 2ab$ | |
| **13** $12fg + 5fh$ | | **14** $4ab + 7ac$ | |
| **15** $10a - 5ab + 3b$ | | **16** $4a + b$ | |
| **17** $3a + 3$ | | **18** $2x + 3y$ | |
| **19** $x - 9y$ | | **20** $5f + 1$ | |
| **21** $yw - xw$ | | **22** $a + 25$ | |
| **23** $5c + 2d$ | | **24** $2x - y$ | |
| **25** $2x + 3$ | | **26** $17 - 6x$ | |
| **27** $1 - 8x$ | | **28** $7a - a^2$ | |
| **29** $10w - 2w^2$ | | **30** $5p - 6p^2$ | |

## R EVISION

### Exercise C

**1**
(a) $2a + 5b$  (b) $2x + 2y$
(c) $x^2 + 4x^3$  (d) $2abc + 5bc + ad$
(e) $12ab$  (f) $9d^2$
(g) $st^2u$  (h) $c^5d^2$
(i) $10a^2b^2c$  (j) $15x^2y^3$
(k) $y^3$  (l) $5q$
(m) $\dfrac{3s}{t}$  (n) $\dfrac{3p^2q^2}{2}$

**2**
(a) $a^2 + ab - b^2$  (b) $9b - 10$
(c) $3ab - 2ad$  (d) $3t - 2$
(e) $15 - 2x$  (f) $7f - f^2$

### Exercise CC

**1**
(a) 23, 27  (b) 31, 36
(c) 19, 30  (d) 12, 10

**2**
(a) $4x + 1$; 41, 61  (b) $3x - 2$; 43, 58
(c) $x^2 + 1$; 101, 225  (d) $2x^2 + 2$; 290, 452

**3**
(a) $3n + 3$  (b) (i) $2n + 1$ (ii) $4n + 4$
(c) $\frac{1}{2}(n^2 + n)$

## 36 FINDING POINTS IN ALL FOUR QUADRANTS

### Exercise 36A

**1** A = (−4, 1), B = (2, 4), C = (2, −3),
D = (−3, −4), E = (−1, 4), F = (0, 1)

**2** H = (−4, 5), I = (2, 1), J = (−3, −3), K = (−3, 0),
L = (5, 3), M = (4, −4)

**3** Q = (4, 5), R = (0, 2), S = (−3, −3), T = (−4, 5),
U = (2, −4), V = (5, 2)

**4** A = (−4, 3), B = (−1, −2), C = (3, −4), D = (4, 4),
E = (−5, 0), F = (5, 1), G = (−5, −5), H = (0, 2)

**5** I = (1, 5), J = (−4, −3), K = (4, −4), L = (−5, 2),
M = (0, −2), N = (5, 1), P = (2, 0), Q = (−2, −5)

**6** R = (−3, 5), S = (−2, 0), T = (−5, −2), U = (4, 4),
V = (4, −4), W = (−1, −4), X = (5, 0), Y = (2, 1)

### Exercise 36B

**1** A = (0, 5), B = (−4, −3), C = (0, −3), D = (5, 1),
E = (−4, 3), F = (4, −5), G = (1, 2), H = (4, 5)

**2** I = (−4, −4), J = (−5, 0), K = (1, −5), L = (5, −3),
M = (−2, 3), N = (2, 1), P = (4, 2), Q = (0, 4)

**3** R = (2, 1), S = (4, −1), T = (−5, 1), U = (1, −4),
V = (−3, 5), W = (−3, 0), X = (3, 4), Y = (−4, −3)

**4** A = (−5, 5), B = (1, −1), C = (5, 3), D = (−3, 0),
E = (4, −4), F = (5, −1), G = (−4, −3), H = (0, 4)

**5** I = (3, 4), J = (−4, −3), K = (3, 0), L = (−2, 2),
M = (−5, 3), N = (1, −3), P = (−4, 0), Q = (5, −5)

**6** R = (−4, 4), S = (−4, −1), T = (0, −4), U = (5, 1),
V = (4, 4), W = (−5, −4), X = (4, −3), Y = (−1, 3)

# 37  PLOTTING POINTS IN ALL FOUR QUADRANTS

## Exercise 37A

**1**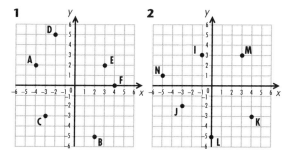

**2**

**3**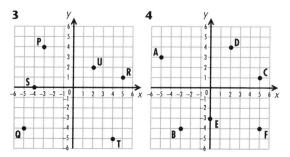

**4**

**5**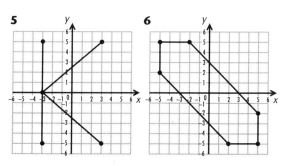

**6**

**7**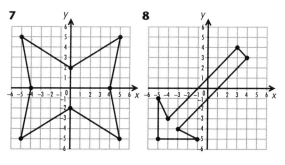

**8**

## Exercise 37B

**1**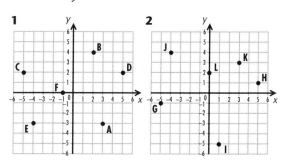

**2**

**3**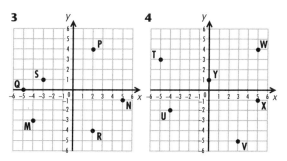

**4**

**5**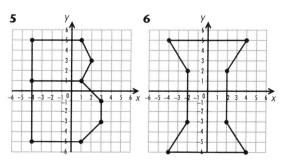

**6**

**7**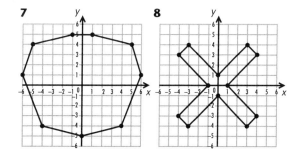

**8**

## 38 DRAWING GRAPHS

### Exercise 38A

**1**

| x | −3 | −2 | −1 | 0 | 1 | 2 | 3 |
|---|---|---|---|---|---|---|---|
| y = 2x | −6 | −4 | −2 | 0 | 2 | 4 | 6 |

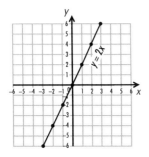

**2**

| x | −3 | −2 | −1 | 0 | 1 | 2 | 3 |
|---|---|---|---|---|---|---|---|
| y = x + 2 | −1 | 0 | 1 | 2 | 3 | 4 | 5 |

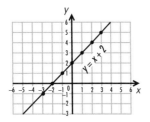

**3**

| x | −2 | −1 | 0 | 1 | 2 | 3 | 4 |
|---|---|---|---|---|---|---|---|
| y = 2x − 2 | −6 | −4 | −2 | 0 | 2 | 4 | 6 |

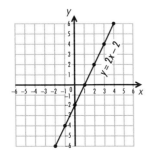

**4**

| x | −2 | −1 | 0 | 1 | 2 | 3 | 4 |
|---|---|---|---|---|---|---|---|
| y = 4x − 5 | −13 | −9 | −5 | −1 | 3 | 7 | 11 |

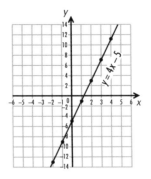

**5**

| x | −2 | −1 | 0 | 1 | 2 | 3 | 4 |
|---|---|---|---|---|---|---|---|
| y = 2x + 3 | −1 | 1 | 3 | 5 | 7 | 9 | 11 |

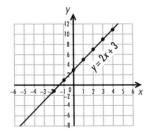

**6**

| x | −2 | −1 | 0 | 1 | 2 | 3 | 4 |
|---|---|---|---|---|---|---|---|
| y = 3x − 4 | −10 | −7 | −4 | −1 | 2 | 5 | 8 |

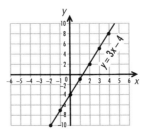

**7**

| x | −3 | −2 | −1 | 0 | 1 | 2 | 3 |
|---|---|---|---|---|---|---|---|
| y = 7 − x | 10 | 9 | 8 | 7 | 6 | 5 | 4 |

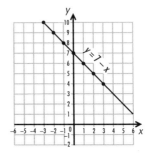

**8**

| $x$ | −3 | −2 | −1 | 0 | 1 | 2 | 3 |
|---|---|---|---|---|---|---|---|
| $y = -3x$ | 9 | 6 | 3 | 0 | −3 | −6 | −9 |

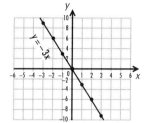

**9**

| $x$ | −2 | −1 | 0 | 1 | 2 | 3 | 4 |
|---|---|---|---|---|---|---|---|
| $y = 3x - 3$ | −9 | −6 | −3 | 0 | 3 | 6 | 9 |

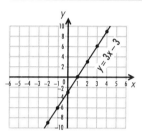

**10**

| $x$ | −1 | 0 | 1 | 2 | 3 | 4 | 5 |
|---|---|---|---|---|---|---|---|
| $y = 2x - 4$ | −6 | −4 | −2 | 0 | 2 | 4 | 6 |

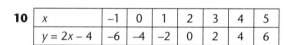

**11**

| $x$ | −3 | −2 | −1 | 0 | 1 | 2 | 3 |
|---|---|---|---|---|---|---|---|
| $y = 2x - 1$ | −7 | −5 | −3 | −1 | 1 | 3 | 5 |

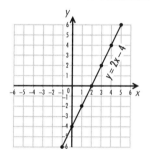

**12**

| $x$ | −2 | −1 | 0 | 1 | 2 | 3 | 4 |
|---|---|---|---|---|---|---|---|
| $y = 5 - x$ | 7 | 6 | 5 | 4 | 3 | 2 | 1 |

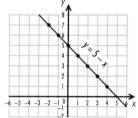

**13**

| $x$ | −4 | −2 | 0 | 2 | 4 | 6 | 8 |
|---|---|---|---|---|---|---|---|
| $y = \frac{1}{2}x - 4$ | −6 | −5 | −4 | −3 | −2 | −1 | 0 |

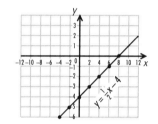

**14**

| $x$ | −1 | 0 | 1 | 2 | 3 | 4 | 5 |
|---|---|---|---|---|---|---|---|
| $y = 10 - 2x$ | 12 | 10 | 8 | 6 | 4 | 2 | 0 |

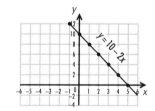

**15**

| $x$ | −6 | −4 | −2 | 0 | 2 | 4 | 6 |
|---|---|---|---|---|---|---|---|
| $y = \frac{1}{2}x + 3$ | 0 | 1 | 2 | 3 | 4 | 5 | 6 |

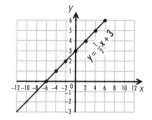

**Exercise** 38B

**1**

| $x$ | –3 | –2 | –1 | 0 | 1 | 2 | 3 |
|---|---|---|---|---|---|---|---|
| $y = 4x$ | –12 | –8 | –4 | 0 | 4 | 8 | 12 |

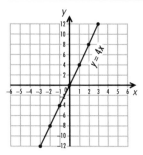

**2**

| $x$ | –3 | –2 | –1 | 0 | 1 | 2 | 3 |
|---|---|---|---|---|---|---|---|
| $y = 2x - 1$ | –7 | –5 | –3 | –1 | 1 | 3 | 5 |

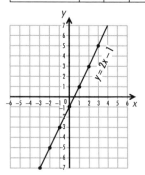

**3**

| $x$ | –2 | –1 | 0 | 1 | 2 | 3 | 4 |
|---|---|---|---|---|---|---|---|
| $y = x - 3$ | –5 | –4 | –3 | –2 | –1 | 0 | 1 |

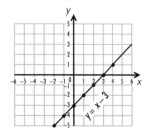

**4**

| $x$ | –3 | –2 | –1 | 0 | 1 | 2 | 3 |
|---|---|---|---|---|---|---|---|
| $y = -2x$ | 6 | 4 | 2 | 0 | –2 | –4 | –6 |

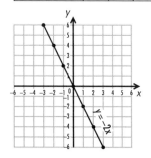

**5**

| $x$ | –2 | –1 | 0 | 1 | 2 | 3 | 4 |
|---|---|---|---|---|---|---|---|
| $y = 3x - 5$ | –11 | –8 | –5 | –2 | 1 | 4 | 7 |

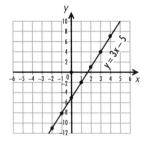

**6**

| $x$ | –3 | –2 | –1 | 0 | 1 | 2 |
|---|---|---|---|---|---|---|
| $y = 3x + 3$ | –6 | –3 | 0 | 3 | 6 | 9 |

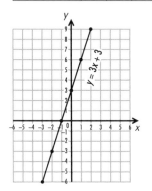

**7**

| $x$ | –3 | –2 | –1 | 0 | 1 | 2 | 3 |
|---|---|---|---|---|---|---|---|
| $y = -x$ | 3 | 2 | 1 | 0 | –1 | –2 | –3 |

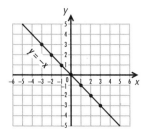

**8**

| $x$ | –3 | –2 | –1 | 0 | 1 | 2 |
|---|---|---|---|---|---|---|
| $y = 2x + 6$ | 0 | 2 | 4 | 6 | 8 | 10 |

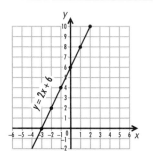

**9**

| x | −2 | −1 | 0 | 1 | 2 | 3 | 4 |
|---|----|----|---|---|---|---|---|
| y = 8 − x | 10 | 9 | 8 | 7 | 6 | 5 | 4 |

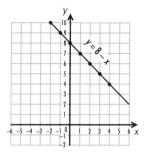

**10**

| x | −2 | −1 | 0 | 1 | 2 | 3 | 4 |
|---|----|----|---|---|---|---|---|
| y = 3x − 2 | −8 | −5 | −2 | 1 | 4 | 7 | 10 |

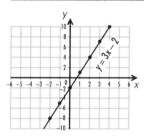

**11**

| x | −2 | −1 | 0 | 1 | 2 | 3 |
|---|----|----|---|---|---|---|
| y = 4x − 2 | −10 | −6 | −2 | 2 | 6 | 10 |

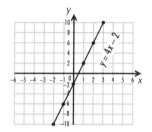

**12**

| x | −2 | −1 | 0 | 1 | 2 | 3 | 4 |
|---|----|----|---|---|---|---|---|
| y = 8 − 2x | 12 | 10 | 8 | 6 | 4 | 2 | 0 |

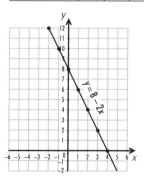

**13**

| x | −6 | −4 | −2 | 0 | 2 | 4 | 6 |
|---|----|----|----|---|---|---|---|
| y = $\frac{1}{2}$x + 1 | −2 | −1 | 0 | 1 | 2 | 3 | 4 |

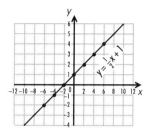

**14**

| x | −1 | 0 | 1 | 2 | 3 | 4 | 5 |
|---|----|---|---|---|---|---|---|
| y = 7 − 3x | 10 | 7 | 4 | 1 | −2 | −5 | −8 |

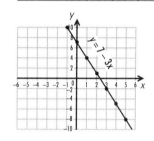

**15**

| x | −4 | −2 | 0 | 2 | 4 | 6 | 8 |
|---|----|----|---|---|---|---|---|
| y = $\frac{1}{2}$x − 2 | −4 | −3 | −2 | −1 | 0 | 1 | 2 |

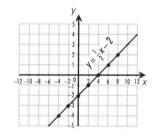

# 39 SOLVING EQUATIONS

**Exercise** 39A

| | | | | |
|---|---|---|---|---|
| **1** 3 | **2** 2 | **3** 5 | **4** 6 | **5** 4 |
| **6** 1 | **7** 9 | **8** 3 | **9** 4 | **10** 1$\frac{1}{2}$ |
| **11** 5 | **12** 2$\frac{1}{2}$ | **13** 6 | **14** 7$\frac{1}{2}$ | **15** 9 |
| **16** 5 | **17** 4 | **18** 3 | **19** 7 | **20** 1$\frac{1}{2}$ |
| **21** 4 | **22** 2 | **23** 4 | **24** 2 | **25** 3 |
| **26** 6 | **27** 1$\frac{1}{2}$ | **28** 5$\frac{1}{2}$ | **29** 7 | **30** 3$\frac{1}{2}$ |

**Exercise** 39B

| | | | | |
|---|---|---|---|---|
| **1** 4 | **2** 1 | **3** 6 | **4** 2 | **5** 7 |
| **6** 8 | **7** 3 | **8** 5$\frac{1}{2}$ | **9** 4 | **10** 2$\frac{1}{2}$ |

| | | | | | | | | | |
|---|---|---|---|---|---|---|---|---|---|
| **11** 4 | **12** 1 | **13** 5 | **14** $3\frac{1}{2}$ | **15** 4 |
| **16** 4 | **17** 1 | **18** 2 | **19** 6 | **20** 5 |
| **21** 3 | **22** 7 | **23** 4 | **24** $5\frac{1}{2}$ | **25** $1\frac{1}{2}$ |
| **26** $2\frac{1}{2}$ | **27** 8 | **28** 3 | **29** 6 | **30** $2\frac{1}{2}$ |

## Exercise 39C

| | | | | |
|---|---|---|---|---|
| **1** 2 | **2** 7 | **3** 9 | **4** 8 | **5** 2 |
| **6** 6 | **7** 3 | **8** 4 | **9** 2 | **10** 5 |
| **11** 8 | **12** 2 | **13** 6 | **14** 3 | **15** 6 |
| **16** 4 | **17** 13 | **18** 5 | **19** 11 | **20** $4\frac{1}{2}$ |
| **21** 9 | **22** 8 | **23** 7 | **24** $1\frac{3}{4}$ | **25** 7 |
| **26** $2\frac{1}{2}$ | **27** 9 | **28** 8 | **29** $4\frac{1}{4}$ | **30** 5 |

## Exercise 39D

| | | | | |
|---|---|---|---|---|
| **1** 3 | **2** 5 | **3** 6 | **4** 2 | **5** 4 |
| **6** 3 | **7** 7 | **8** 16 | **9** 4 | **10** 2 |
| **11** 5 | **12** 3 | **13** $2\frac{1}{2}$ | **14** 8 | **15** 2 |
| **16** 7 | **17** 3 | **18** 9 | **19** 8 | **20** 5 |
| **21** $3\frac{1}{4}$ | **22** 6 | **23** 12 | **24** $1\frac{1}{4}$ | **25** $1\frac{1}{2}$ |
| **26** 15 | **27** 9 | **28** $2\frac{3}{4}$ | **29** 8 | **30** 7 |

# 40 WRITING EQUATIONS

## Exercise 40A

**1** $2x + 9 = 17$; 4
**2** $x + 9 = 14$; 5
**3** $3(x + 5) = 19$; $1\frac{1}{3}$
**4** $4x - 5 = 19$; 6
**5** $8x - 6 = 38$; $5\frac{1}{2}$
**6** $7x - 9 = 40$; 7
**7** $3(x - 4) = 24$; 12
**8** $9x - 7 = 29$; 4
**9** $2x + 20 = 6x - 4$; 6
**10** $5(x - 2) = 3(x + 6)$; 14
**11** $x + x + 5 = 31$; 13
**12** $x + x + x + 2x = 15$; £3
**13** $3x + x = 18$; £4.50
**14** $94 - 3x = 28$; 22 cm
**15** $x + x + 1 = 93$; 46, 47
**16** $x + 3x + 4x = 32$; £4
**17** $x + 3 + x + x + 3 - 5 = 64$; Ali 21, Ann 24, Jomo 19
**18** $x + x - 6 + x - 14 = 37$; 19 years old
**19** $x + 4 + x + x + 6 = 16$; 2
**20** $x + 12 + x + x + 15 = 63$; 12
**21** $2x + x + 60 = 180$; 40°
**22** $2x + x + 55 + x - 15 = 180$; 35°
**23** $3x + x + 20 = 180$; 40°

**24** $x - 10 + 2x - 50 + 3x = 360$; 70°
**25** $x + x + 70 + 3x + 40 = 360$; 50°
**26** $x + x + 10 + 50 = 180$; 60°
**27** $x + x - 20 + x - 10 = 180$; 70°
**28** $x + x + x - 30 + 2x - 10 = 360$; 80°
**29** $5(x + 3) = 35$; 4
**30** $6(2x - 1) = 54$; 5

## Exercise 40B

**1** $3x + 5 = 17$; 5
**2** $x - 12 = 13$; 25
**3** $6x + 18 = 39$; $3\frac{1}{2}$
**4** $5x + 3 = 48$; 9
**5** $7x + 2 = 37$; 5
**6** $3x - 2 = 25$; 9
**7** $6x - 7 = 29$; 6
**8** $4x + 5 = 29$; 6
**9** $5x - 6 = 24$; 6
**10** $2(x - 6) = \frac{x}{2}$; 8
**11** $x + x + 3 = 25$; 11
**12** $x + x + x - 3 + x - 3 = 42$; 12
**13** $x + 3 + x = 27$; 12
**14** $4x + x = 65$; 13
**15** $x + x + 1 = 93$; 46, 47
**16** $x + x + \frac{1}{2}x = 30$; $x = 12$, short side = 6
**17** $x - 5 = 26$; Martin = 31, Mark = 8
**18** $3x - 9 = 18$; 9
**19** $x + x + 3 + x - 1 = 29$; 9
**20** $x + x + 4 + x - 3 = 22$; 7
**21** $x + x - 40 = 180$; 110°
**22** $2x - 50 + 3x = 180$; 46°
**23** $2x + x + 45 + 2x - 50 = 180$; 37°
**24** $2x + 2x - 30 + 2x + 30 = 360$; 60°
**25** $x + 50 + 4x + 3x + 30 = 360$; 35°
**26** $2x + 10 + 2x + x + 2x = 360$; 50°
**27** $x + x + 20 + x - 20 = 180$; 60°
**28** $x - 15 + 2x + 3x = 180$; $32\frac{1}{2}$°
**29** $4(3x + 2) = 20$; 1
**30** $7(x - 4) = 14$; 6

# 41 TRIAL AND IMPROVEMENT

## Exercise 41A

| | |
|---|---|
| **1** 1.5, –1.6 | **2** –1.62, 0.62 |
| **3** –1.45, 3.45 | **4** 1.00, –0.67 |
| **5** –0.86, 1.36 | **6** –1.54, 0.87 |
| **7** 1.91 | **8** –2.57, 0.91 |
| **9** 2.78, 0.72 | **10** –4.12, 0.12 |
| **11** 2.29, –0.29 | **12** 7.74, 0.26 |

**13** 1.68, 0.12

**15** −4.24, 0.24

**17** 5.65, 0.35

**19** 2.35, −0.85

**14** 2.47

**16** 2.29

**18** 3.00

**20** −4.23, 0.23

### Exercise 41B

**1** 1.5, −0.75

**3** −0.88, 0.68

**5** 7.65, 2.35

**7** −0.80, 2.80

**9** −3.16, 0.16

**11** 5.65, 0.35

**13** 2.67

**15** −5.45, −0.55

**17** 3.17

**19** 6.24, −0.24

**2** −3.45, 1.45

**4** −0.62, 1.62

**6** 2.08

**8** 3.30

**10** −2.35, 0.85

**12** 4.65, −0.65

**14** 3.69, −0.36

**16** −1.5 only

**18** −6.70, −0.30

**20** 7.69, −0.19

## 42 SOLVING SIMULTANEOUS EQUATIONS BY ALGEBRAIC METHODS

### Exercise 42A

**1** $x = 4, y = 2$

**3** $x = 6, y = 3$

**5** $x = 2, y = 2$

**7** $x = 4, y = 1$

**9** $x = 5, y = 1$

**11** $x = 2, y = 1$

**13** $x = 4, y = -2$

**15** $x = 2, y = 0$

**17** $x = 1, y = 2$

**19** $x = 5, y = 2$

**21** $x = 2, y = 1$

**23** $x = \frac{1}{3}, y = -\frac{1}{2}$

**25** $x = 3, y = -2$

**27** $x = 1, y = 2$

**29** $x = 3, y = \frac{1}{2}$

**2** $x = 5, y = 1$

**4** $x = 1, y = 2$

**6** $x = 2, y = 3$

**8** $x = 7, y = 2$

**10** $x = 7, y = 3$

**12** $x = 5, y = 1$

**14** $x = 2, y = 1$

**16** $x = 1, y = 1$

**18** $x = 1, y = 1$

**20** $x = 2, y = -2$

**22** $x = 3, y = 1$

**24** $x = 3, y = 2$

**26** $x = 4, y = 1$

**28** $x = 1, y = \frac{1}{3}$

**30** $x = \frac{1}{3}, y = -\frac{1}{2}$

### Exercise 42B

**1** $x = 7, y = 6$

**3** $x = 4, y = 3$

**5** $x = 3, y = 1$

**7** $x = 8, y = 3$

**9** $x = 4, y = 2$

**11** $x = 1, y = 2$

**13** $x = 1, y = 7$

**15** $x = 1, y = 1$

**17** $x = \frac{1}{2}, y = 1$

**19** $x = 0, y = -1$

**21** $x = 1, y = 2$

**23** $x = 1, y = \frac{1}{2}$

**25** $x = 4, y = 3$

**2** $x = 8, y = 3$

**4** $x = 2, y = 3$

**6** $x = 3, y = 1$

**8** $x = 9, y = 4$

**10** $x = 5, y = 2$

**12** $x = 5, y = 3$

**14** $x = 2, y = 5$

**16** $x = 4, y = 1$

**18** $x = 6, y = -1$

**20** $x = 3, y = -3$

**22** $x = 3, y = 1$

**24** $x = 3, y = 2$

**26** $x = 3, y = 0$

**27** $x = 1, y = -2$

**29** $x = 0, y = -3$

**28** $x = \frac{1}{2}, y = 1$

**30** $x = 3, y = \frac{1}{2}$

### Exercise 42C

**1** Pen 10p, ruler 15p

**2** £2

**3** 6.3 kg, 4.2 kg

**4** Apple 50 g, orange 25 g

**5** £1.20

**6** 1.2 litre, 2.2 litres

**7** 9, 18

**8** 2, 3

**9** 60p, 25p

**10** £15, £20

**11** 3 kg, $1\frac{1}{2}$ kg

**12** 45p, 30p

**13** 2p, 1p

**14** 1 pint, $\frac{1}{2}$ pint

**15** 4 m, 1 m

### Exercise 42D

**1** Motorbike £2400, mountain bike £1600

**2** Mechanics £160, assistants £100

**3** 6p, 8p

**4** 9, 3

**5** 2.5 cm, 2 cm

**6** £1, £6

**7** 40p, 20p

**8** 6 hours, 2 hours

**9** £240, £100

**10** 20, 40

**11** £1, £3

**12** 30p, 20p

**13** 1 pint, $\frac{1}{2}$ pint

**14** 3 lb, $\frac{1}{4}$ lb

**15** 2 ft, 6 ft

## 43 SOLVING SIMULTANEOUS EQUATIONS BY GRAPHICAL METHODS

### Exercise 43A

**1** $x = 5, y = 3$

**3** $x = 4, y = 4$

**5** $x = 1, y = 6$

**7** $x = 6, y = 4$

**9** $x = -2, y = -2$

**11** $x = 2, y = 3$

**13** $x = 4, y = 5$

**15** $x = 3, y = 5$

**17** $x = -2, y = 3$

**19** $x = 4, y = 6$

**2** $x = 3, y = 5$

**4** $x = 4, y = 5$

**6** $x = 3, y = 2$

**8** $x = -3, y = 4$

**10** $x = 4, y = -2$

**12** $x = 2, y = 3$

**14** $x = 4, y = 6$

**16** $x = 2\frac{1}{2}, y = 3\frac{1}{2}$

**18** $x = -2, y = -3$

**20** $x = -2, y = -1$

### Exercise 43B

**1** $x = 6, y = 2$

**3** $x = 7, y = 1$

**5** $x = 4, y = 3$

**7** $x = 5, y = 7$

**9** $x = -1, y = 3$

**11** $x = 2, y = 5$

**13** $x = \frac{1}{2}, y = 2\frac{1}{2}$

**15** $x = 4\frac{1}{2}, y = 3\frac{1}{2}$

**17** $x = -1, y = 1$

**19** $x = 2, y = -1$

**2** $x = 3, y = 4$

**4** $x = 5, y = 5$

**6** $x = 6, y = 5$

**8** $x = -3, y = -4$

**10** $x = 2, y = -3$

**12** $x = 2, y = 3$

**14** $x = 2, y = 5$

**16** $x = 1\frac{1}{2}, y = 2$

**18** $x = 3, y = -1$

**20** $x = 1, y = 3\frac{1}{2}$

# Exercise 43C

**1** $x = 4, y = -2$      **2** $x = -4, y = -5$
**3** $x = 4, y = 0$      **4** $x = 3, y = 6$
**5** $x = 2, y = 5$      **6** $x = -4, y = 3$
**7** $x = 2, y = 5$      **8** $x = 1, y = \frac{1}{2}$
**9** $x = 1, y = 2$      **10** $x = 1, y = 3$
**11** $x = 3, y = 0$      **12** $x = 3, y = 4$
**13** $x = 5, y = 2$      **14** $x = 3, y = 5$
**15** $x = 3, y = 6$

# Exercise 43D

**1** $x = 1, y = -2$      **2** $x = 2, y = 0$
**3** $x = 3, y = 5\frac{1}{2}$      **4** $x = 6, y = 4$
**5** $x = 5, y = 9$      **6** $x = -3, y = -2\frac{1}{2}$
**7** $x = 2, y = 2$      **8** $x = 0, y = 6$
**9** $x = 2, y = 3$      **10** $x = 4, y = 3$
**11** $x = 3, y = 1$      **12** $x = 5, y = 1$
**13** $x = 1, y = 7$      **14** $x = \frac{1}{2}, y = 0$
**15** $x = -3, y = -1$

# 44 INEQUALITIES

# Exercise 44A

**1** $x < 1$     **2** $x \geq -4$     **3** $x \leq 5$
**4** $x > 0$     **5** $x < 2$     **6** $x > 5$
**7** $x \geq -1$     **8** $x \leq -2$     **9** $x < -3$
**10** $x < 4$     **11** $x > 5$     **12** $x \geq 5$
**13** $1 < x \leq 4$     **14** $-2 < x < 3$     **15** $-3 \leq x < -1$

**16** 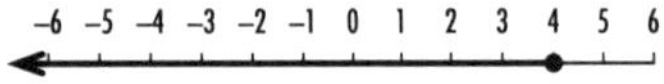
**17** 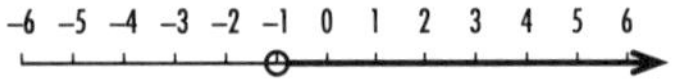
**18** 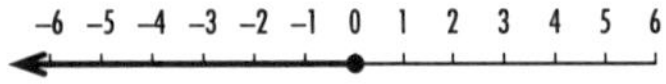
**19** 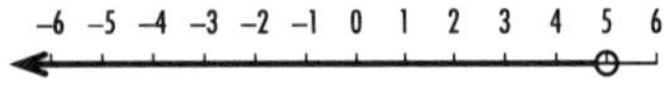
**20** 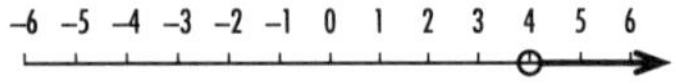
**21** 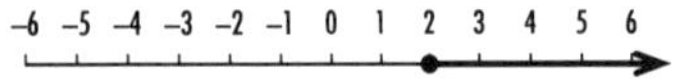
**22** 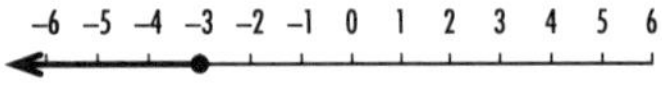
**23** 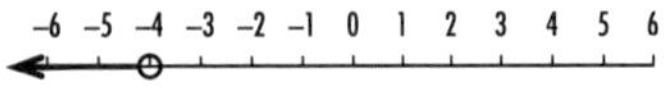
**24** 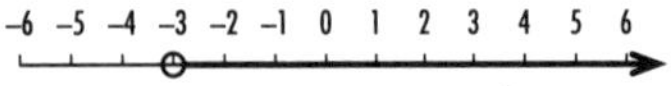
**25** 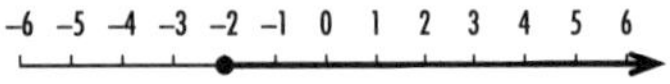
**26**

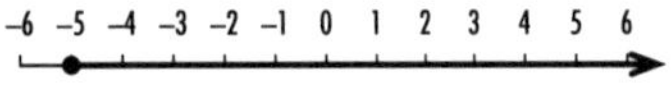

**27** 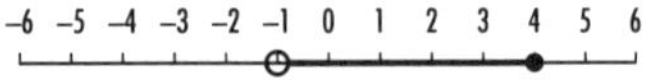
**28** 
**29** 
**30**

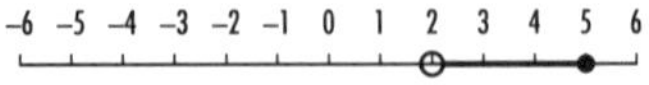

# Exercise 44B

**1** $x < -1$     **2** $x > -5$     **3** $x \leq 2$
**4** $x \geq 1$     **5** $x \geq -3$     **6** $x \geq 4$
**7** $x \leq -4$     **8** $x < 3$     **9** $x < -2$
**10** $x < -5$     **11** $x \geq 3$     **12** $-2 < x < 1$
**13** $-1 \leq x < 2$     **14** $-3 \leq x \leq 1$     **15** $-4 < x < -1$

**16** 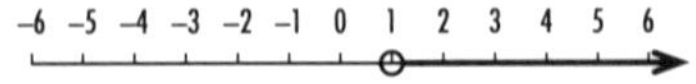
**17** 
**18** 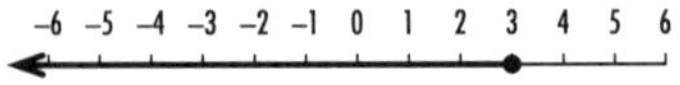
**19** 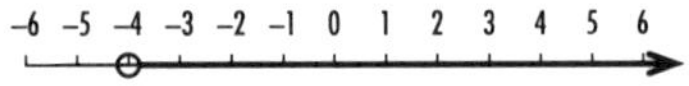
**20** 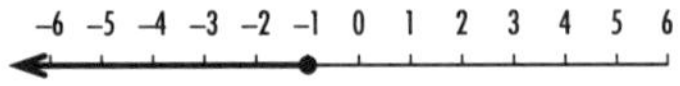
**21** 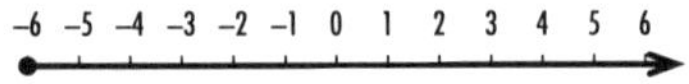
**22** 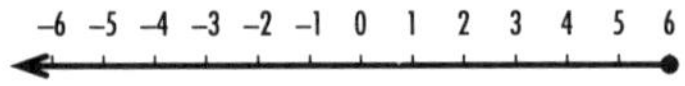
**23** 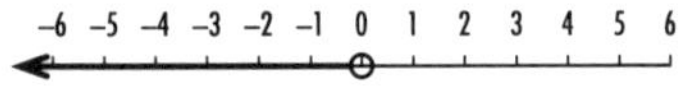
**24** 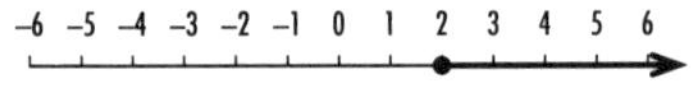
**25** 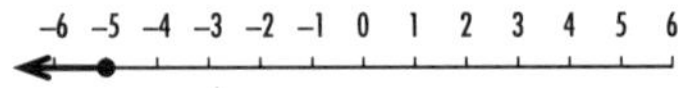
**26** 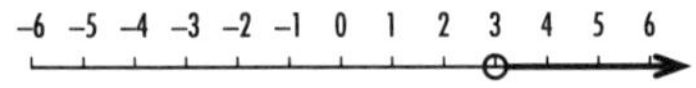
**27** 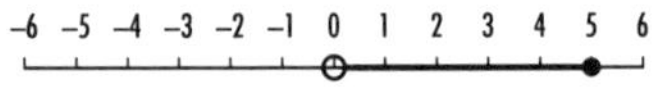
**28** 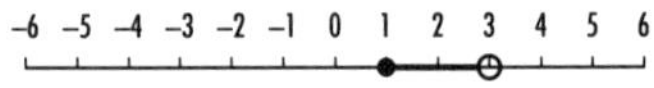
**29** 
**30**

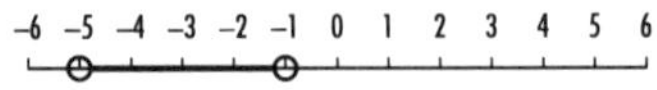

# Exercise 44C

**1** $-3, -2, ..., 4, 5$      **2** $-5, -4, ..., 6, 7$
**3** $0, 1, 2, 3, 4, 5, 6$      **4** $-5, -4, -3, -2, -1$
**5** $-8, -7, -6, -5, -4, -3, -2, -1$
**6** $5, 6, 7, 8$      **7** $0, 1, 2, 3, 4, 5, 6, 7$
**8** $-3, -2, -1, 0, 1$      **9** $-8, -6, -4, -2, 0, 2$

**10** –2, 0, 2

**11** 0, 2, 4, 6

**12** –8, –6, –4, –2

**13** –6, –4, –2, 0, 2

**14** 6

**15** 1, 3, 5, 7, 9

**16** –1, 1, 3, 5

**17** –3, –1, 1

**18** –3, –1, 1, 3

**19** –7, –5, –3, –1

**20** –1, 1, 3, 5

## Exercise 44D

**1** –4, –3, –2, –1, 0, 1, 2

**2** 0, 1, 2, 3, 4, 5

**3** 0, 1, 2, 3, 4

**4** –1, 0, 1, 2, 3, 4

**5** –6, –5, –4, –3, –2

**6** –2, –1, 0, 1, 2, 3

**7** –3, –2, –1, 0, 1, 2

**8** –6, –5, –4, –3, –2

**9** –4, –2, 0, 2, 4

**10** –2, 0, 2, 4, 6

**11** –6, –4, –2, 0

**12** –4, –2, 0, 2

**13** 0, 2, 4, 6

**14** –6, –4

**15** –3, –1, 1, 3, 5

**16** –1, 1, 3, 5, 7

**17** 3, 5, 7, 9

**18** –7, –5, –3, –1, 1

**19** –3, –1, 1, 3, 5

**20** 1, 3, 5

## $R$EVISION

## Exercise $D$

**1** A(5, 5), B(–4, –4), C(–2, –1), D(2, –5), E(–2, 5), F(–4, 2), G(0, 3), H(5, –2), I(0, –4), J(2, 0), K(4, 1)

**2**

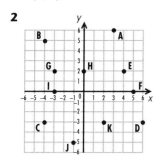

**3**

| $x$ | –2 | –1 | 0 | 1 | 2 | 3 | 4 |
|---|---|---|---|---|---|---|---|
| $y$ | –7 | –5 | –3 | –1 | 1 | 3 | 5 |

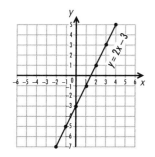

**4**

| $x$ | –6 | –4 | –2 | 0 | 2 | 4 | 6 |
|---|---|---|---|---|---|---|---|
| $y$ | –1 | 0 | 1 | 2 | 3 | 4 | 5 |

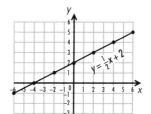

**5** (a) 3  (b) 5  (c) 4  (d) $1\frac{1}{2}$  (e) $2\frac{1}{2}$  (f) 2
  (g) 7  (h) $4\frac{1}{2}$  (i) $1\frac{2}{3}$  (j) $4\frac{1}{2}$  (k) 9  (l) $3\frac{1}{4}$

**6** (a) +7.07, –7.07  (b) 3.29, –3.79
  (c) 3.91  (d) 2.57
  (e) 18.34, –16.34  (f) 2.35

**7** (a) $x = 5, y = 1$  (b) $x = 10, y = 1$
  (c) $x = 1, y = 2$  (d) $x = 10, y = 3$
  (e) $x = 7, y = 1$  (f) $x = 2, y = 8$
  (g) $x = 1, y = 3$  (h) $x = 5, y = 5$

**8** (a)

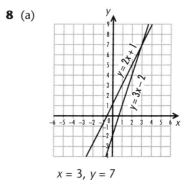

$x = 3, y = 7$

(b)

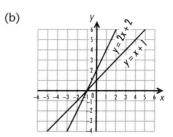

$x = –1, y = 0$

**9** (a) –2, –1, 0, 1, 2, 3  (b) –5, –4, –3, –2
  (c) 5, 6, 7, 8  (d) –2, –1, 0, 1, 2

**10** (a) –2, 0, 2, 4, 6  (b) 0, 2, 4, 6
  (c) –6, –4, –2, 0  (d) 0, 2, 4

**Exercise** $\triangleright\triangleright$

**1** (a) $2x + 3 = 5$; $x = 1$     (b) $x - 6 = 3$; $x = 9$
    (c) $3x - 1 = 11$; $x = 4$     (d) $7x + 4 = 25$; $x = 3$
    (e) $8x - 3 = 9$; $x = 1\frac{1}{2}$     (f) $2x + 4 = 28$; $x = 12$
    (g) $2x - 5 = 25$; $x = 15$

**2** $4x + 6 = 22$; $x = 4$

**3** (a) $3x + x + 50 + 2x + 10 = 360$, $50°$

    (b) $2x + 2x + 30 + x + 2x + 15 = 360$, $45°$

**4** (a) 10p           (b) 7p

**5** (a) 5 g           (b) 8 g

**6** (a) 6 $l$           (b) 2 $l$

**7** (a) 6 cm        (b) 4 cm

**8** (a) £16         (b) £20

# Shape, space and measures

## 45 FINDING UNKNOWN ANGLES ON PARALLEL LINES GIVING REASONS

### Exercise 45A

1. $a = 52°$ (vertically opposite angles),
   $b = 52°$ (corresponding angles)
2. $c = 37°$ (alternate angles),
   $d = 143°$ (supplementary angles),
   $e = 37°$ (corresponding angles)
3. $f = 48°$ (vertically opposite angles),
   $g = 132°$ (supplementary angles),
   $h = 48°$ (corresponding angles)
4. $i = 111°$ (corresponding angles),
   $j = 111°$ (vertically opposite angles)
5. $k = 63°$ (alternate angles),
   $l = 63°$ (corresponding angles),
   $m = 117°$ (supplementary angles),
   $n = 63°$ (corresponding angles) ,
   $p = 63°$ (vertically opposite angles)
6. $r = 59°$ (corresponding angles),
   $s = 59°$ (vertically opposite angles),
   $t = 88°$ (supplementary angles)
7. $u = 98°$ (vertically opposite angles),
   $v = 78°$ (vertically opposite angles),
   $w = 78°$ (alternate angles),
   $x = 98°$ (corresponding angles),
   $y = 102°$ (supplementary angles)
8. $a = 70°$ (supplementary angles),
   $b = 75°$ (supplementary angles),
   $c = 105°$ (corresponding angles),
   $d = 110°$ (corresponding angles),
   $e = 110°$ (vertically opposite angles),
   $f = 105°$ (vertically opposite angles)
9. $g = 77°$ (corresponding angles),
   $h = 77°$ (alternate angles),
   $i = 77°$ (vertically opposite angles),
   $j = 77°$ (corresponding angles)
10. $k = 75°$ (supplementary angles),
    $l = 115°$ (vertically opposite angles),
    $m = 115°$ (alternate angles),
    $n = 115°$ (corresponding angles),
    $p = 115°$ (corresponding angles)

### Exercise 45B

1. $a = 57°$ (alternate angles),
   $b = 123°$ (supplementary angles),
   $c = 57°$ (corresponding angles)
2. $d = 79°$ (vertically opposite angles),
   $e = 79°$ (corresponding angles)
3. $f = 47°$ (vertically opposite angles),
   $g = 47°$ (alternate angles),
   $h = 47°$ (corresponding angles)
4. $i = 96°$ (alternate angles),
   $j = 96°$ (corresponding angles),
   $k = 96°$ (vertically opposite angles)
5. $l = 63°$ (vertically opposite angles),
   $m = 63°$ (corresponding angles),
   $n = 63°$ (alternate angles),
   $p = 68°$ (supplementary angles),
   $q = 112°$ (alternate angles),
   $r = 112°$ (corresponding angles)
6. $s = 121°$ (vertically opposite angles),
   $t = 121°$ (corresponding angles),
   $u = 121°$ (alternate angles),
   $v = 121°$ (corresponding angles),
   $w = 59°$ (supplementary angles),
   $x = 121°$ (alternate angles)
7. $y = 88°$ (corresponding angles),
   $z = 88°$ (alternate angles),
   $a = 52°$ (alternate angles),
   $b = 52°$ (corresponding angles),
   $c = 52°$ (vertically opposite angles)
8. $d = 62°$ (vertically opposite angles),
   $e = 62°$ (alternate angles),
   $f = 62°$ (corresponding angles),
   $g = 62°$ (corresponding angles),
   $h = 118°$ (supplementary angles)
9. $i = 107°$ (supplementary angles),
   $j = 108°$ (vertically opposite angles),
   $k = 73°$ (corresponding angles),
   $l = 73°$ (vertically opposite angles),
   $m = 108°$ (alternate angles),
   $n = 72°$ (supplementary angles)
10. $p = 89°$ (corresponding angles), $q = 89°$ (corresponding angles), $r = 91°$ (supplementary angles), $s = 89°$ (vertically opposite angles), $t = 89°$ (alternate angles), $u = 91°$ (a number of indirect valid reasons)

# 46 FINDING UNKNOWN ANGLES IN VARIOUS SITUATIONS GIVING REASONS

## Exercise 46A

There are often valid alternative reasons. You may wish to apply follow through after wrong answers.

1  $a = 57°$ (angles of a triangle),
   $b = 123°$ (angles on a straight line)
2  $c = 157°$ (angles at a point)
3  $d = 108°$ (angles on a straight line), $e = 50°$ (angles of a triangle), $f = 50°$ (corresponding angles), $g = 72°$ (corresponding angles)
4  $h = 120°$ (angles of a quadrilateral), $i = 60°$ (angles on a straight line)
5  $j = 90°$ (supplementary angles), $k = 80°$ (alternate angles), $l = 43°$ (angles of a triangle)
6  $m = n = 69°$ (base angles of isosceles triangle), $p = 111°$ (angles on a straight line)
7  $q = 145°$ (angles of a quadrilateral), $r = 86°$ (angles on a straight line), $s = 61°$ (vertically opposite and angles at a point)
8  $t = 39°$ (alternate angles), $u = 105°$ (angles of a triangle and vertically opposite angles), $v = 36°$ (angles of a triangle)
9  $w = 43°$ (angles of a triangle), $x = 47°$ (because $43° + x = 90°$), $y = 16°$ ($90° - 74°$)
10  $a = 129°$ (angles of a a quadrilateral), $b = 116°$ (angles on a straight line), $c = 13°$ (angles of a triangle)
11  $d = 77°$ (angles of a triangle), $e = 103°$ (angles on a straight line), $f = 49°$ (angles of a triangle), $g = 131°$ (angles on a straight line)
12  $h = 63°$ (angles on a straight line), $i = 63°$ (alternate angles), $j = 117°$ (alternate angles), $k = 39°$ (angles at a point)

## Exercise 46B

There are often valid alternative reasons. You may wish to apply follow through after wrong answers.

1  $a = 58°$ (angles of a triangle), $b = 87°$ (angles on a straight line), $c = 125°$ (angles on a straight line)
2  $d = 73°$ (corresponding angles), $e = 55°$ (supplementary angles), $f = 107°$ (supplementary angles), $g = 125°$ (vertically opposite angles)
3  $h = 42°$ (angles of a triangle), $i = 142°$ (angles at a point)
4  $j = 115°$ (angles at a point), $k = 110°$ (angles of a quadrilateral), $l = 70°$ (angles on a straight line)

5  $m = 65°$ (angles of a triangle), $n = 90°$ (supplementary angles), $p = 100°$ (angles of a quadrilateral)
6  $q = 116°$ (angles of a triangle), $r = 64°$ (angles on a straight line), $s = 69°$ (angles of a triangle), $t = 111°$ (angles on a straight line)
7  $u = 109°$ (angles of a quadrilateral), $v = 93°$ (angles on a straight line)
8  $w = 66°$ (alternate angles), $x = 66°$ (vertically opposite angles), $y = 80°$ (angles of a triangle), $a = 80°$ (corresponding angles)
9  $b = 50°$ (angles on a straight line), $c = 18°$ (angles on a straight line)
10  $d = 53°$ (alternate angles), $e = 58°$ (angles of a triangle), $f = 69°$ (alternate angles), $g = 111°$ (angles on a straight line)
11  $h = 60°$ (angles of a triangle), $i = 102°$ (angles on a straight line), $j = 39°$ (angles of a triangle)
12  $k = 60°$ (equilateral triangle), $l = 10°$ (base angles of isosceles triangle)

# 47 THE PROPERTIES OF QUADRILATERALS AND POLYGONS

## Exercise 47A

1  $a = 101°$, $b = 128°$, $c = 79°$
2  $d = 28°$, $e = 111°$, $f = 8$ cm, $g = 4$ cm
3  $h = 31$ mm, $i = 56$ mm, $j = 98°$, $k = 115°$
4  $l = 35$ cm, $m = 72°$, $n = 72°$, $p = 108°$
5  $q = 5$ cm, $r = 68°$, $s = 68°$, $t = 112°$
6  $u = 5$ cm, $v = 4$ cm, $w = 25°$, $x = 100°$
7  $a = b = c = 8$ cm, $d = 90°$
8  $e = 32°$, $f = 58°$, $g = 90°$, $h = 12$ cm
9  $48°$, $132°$, $132°$
10  7, heptagon
11  $122°$, $29°$, $29°$
12

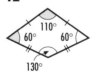  

13  (a) $60°$    (b) $60°$    (c) hexagon
14  Diagonals bisect each other. Diagonals are perpendicular to each other. Diagonals bisect angles.
15  (a) $36°$    (b) $36°$    (c) $144°$

## Exercise 47B

1  $a = 73$ mm, $b = 35$ mm, $c = 90°$, $d = 53°$
2  $e = f = g = 43$ mm, $h = 90°$, $i = 45°$, $j = 45°$

**3** $k = 56°$, $l = 25°$, $m = 118°$
**4** $n = 60°$, $p = 120°$, $q = 60°$
**5** $r = 90°$, $s = 39°$, $t = 41°$, $u = 100°$
**6** $v = w = 7.5$ cm, $x = y = 35°$, $z = 110°$
**7** $a = b = 30°$, $c = 8$ cm, $d = 6$ cm
**8** $e = 4$ m, $f = g = 83°$, $h = 97°$
**9**

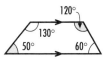 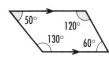

**10** (a) 135°   (b) 45°   (c) octagon
**11** No
**12** 5, pentagon
**13**

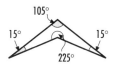

**14** (a) 40°   (b) 40°   (c) 140°
**15**

48 **SYMMETRY INCLUDING PROPERTIES OF QUADRILATERALS AND POLYGONS**

**Exercise** 48A

**1**

**2**
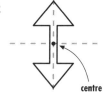

**3**
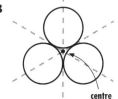

**4**
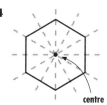

**5**

**6**

**7**
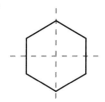

**8**

**9**
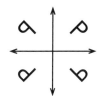

**10**
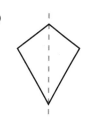

**11**
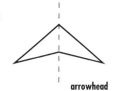

arrowhead

**12**

parallelogram

**13**

centre

**14** 5 of F, G, J, L, P, Q, R and any non-symmetrical representations

**15**

**Exercise** 48B

**1**
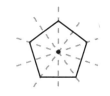

**2**

**3**
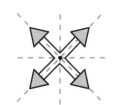

**4**

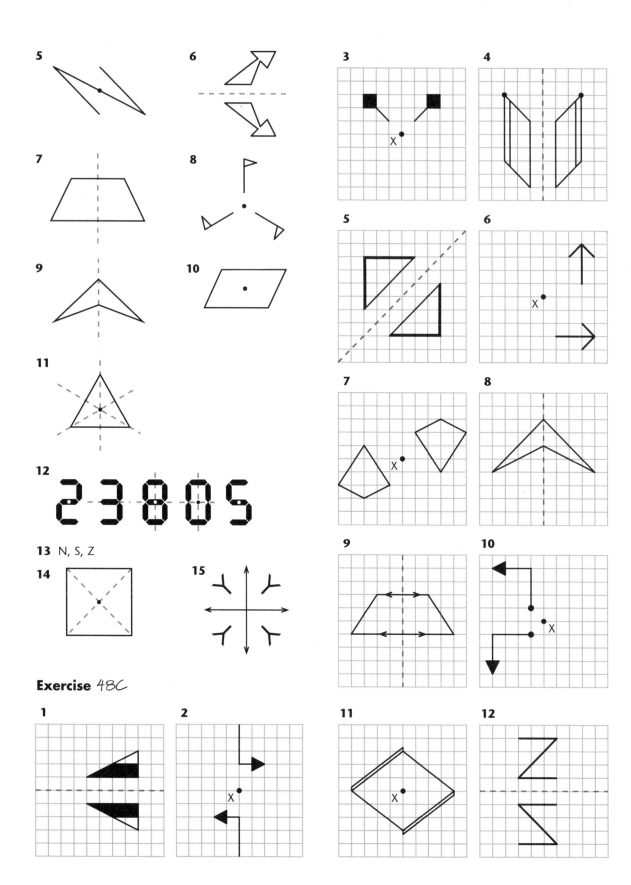

**5** N, S, Z

**Exercise** 48C

## Exercise 48D

**1**

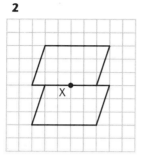

**2**

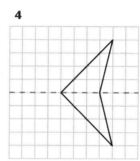

**3**

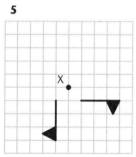

**4**

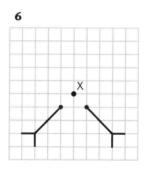

**5**

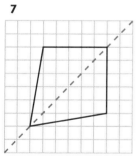

**6**

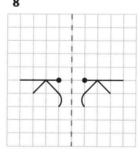

**7**

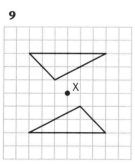

**8**

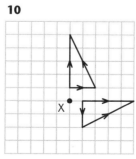

**9**

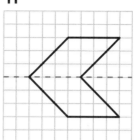

**10**

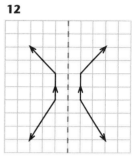

**11**

**12**

# 49 SCALE: CONVERTING LENGTHS

## Exercise 49A

| | | | | | |
|---|---|---|---|---|---|
| **1** 8 km | | **2** 3.5 m | | **3** 70 km | |
| **4** 12 km | | **5** 30 m | | **6** 42 m | |
| **7** 4 m | | **8** 50 m | | **9** 15 km | |
| **10** 2.8 m | | **11** 7 mm | | **12** 8 cm | |
| **13** 6 cm | | **14** 9 cm | | **15** 15 mm | |
| **16** 3 cm | | **17** 12 mm | | **18** 25 mm | |
| **19** 3 mm | | **20** 4 cm | | **21** 10 m | |
| **22** 20 mm | | **23** 16 km | | **24** 5 mm | |
| **25** 9 mm | | **26** 60 km | | **27** 90 km | |
| **28** 45 mm | | **29** 2 cm | | **30** 50 km | |

## Exercise 49B

| | | | | | |
|---|---|---|---|---|---|
| **1** 4 km | | **2** 4 m | | **3** 13 km | |
| **4** 180 m | | **5** 45 km | | **6** 5 m | |
| **7** 11 m | | **8** 3 km | | **9** 6 km | |
| **10** 10 km | | **11** 3 mm | | **12** 2 cm | |
| **13** 15 mm | | **14** 7 cm | | **15** 12 mm | |
| **16** 55 mm | | **17** 13 mm | | **18** 7 cm | |
| **19** 12 mm | | **20** 2 cm | | **21** 20 km | |
| **22** 3 mm | | **23** 17 cm | | **24** 10 m | |
| **25** 24 km | | **26** 25 mm | | **27** 16 mm | |
| **28** 7 km | | **29** 150 km | | **30** 3 cm | |

# 50 DRAWING AND READING TO SCALE

## Exercise 50A

| | |
|---|---|
| **1** 3 cm, 6 km | **2** 5 cm, 25 km |
| **3** 13 mm, 13 km | **4** 4 cm, 20 m |
| **5** 25 mm, 2.5 km | **6** 7 cm, 14 m |
| **7** 45 mm, 45 m | **8** 6 cm, 12 km |
| **9** 28 mm, 2.8 km | **10** 17 mm, 3.4 km |

**11** line of length 6 cm

**12** line of length 3 cm

**13** line of length 25 mm

**14** line of length 25 mm
**15** line of length 75 mm
**16** line of length 11 mm
**17** line of length 14 cm
**18** line of length 27 mm
**19** line of length 24 mm
**20** line of length 10 mm
**21–25** accurate scale drawings

## Exercise 50B

| | |
|---|---|
| **1** 8 cm, 40 km | **2** 9 cm, 18 km |
| **3** 36 mm, 36 m | **4** 5 cm, 10 m |
| **5** 39 mm, 39 m | **6** 3 cm, 150 km |
| **7** 65 mm, 65 km | **8** 2 cm, 100 km |
| **9** 18 mm, 36 m | **10** 6 cm, 300 km |

**11** line of length 3 cm
**12** line of length 70 mm
**13** line of length 11 mm
**14** line of length 13 mm
**15** line of length 36 mm
**16** line of length 85 mm
**17** line of length 11 mm
**18** line of length 70 mm
**19** line of length 13 cm
**20** line of length 23 mm
**21–25** accurate scale drawings

## 51 ENLARGEMENT: SCALE FACTOR AND CENTRE OF ENLARGEMENT

## Exercise 51A

**1**
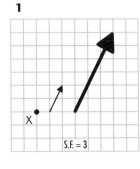
S.F. = 3

**2**
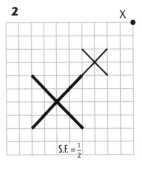
S.F. = ½

**3**
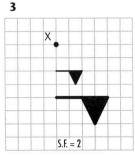
S.F. = 2

**4**
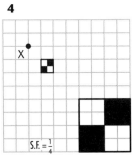
S.F. = ¼

**5**
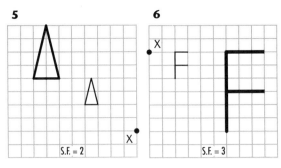
S.F. = 2

**6**
S.F. = 3

**7**
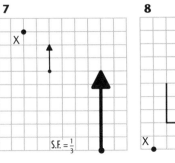
S.F. = ⅓

**8**
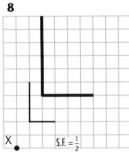
S.F. = ½

**9**
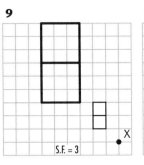
S.F. = 3

**10**
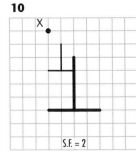
S.F. = 2

## Exercise 51B

**1**
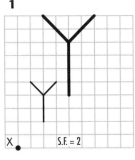
S.F. = 2

**2**
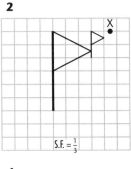
S.F. = ⅓

**3**
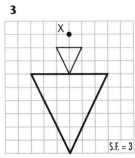
S.F. = 3

**4**
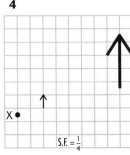
S.F. = ¼

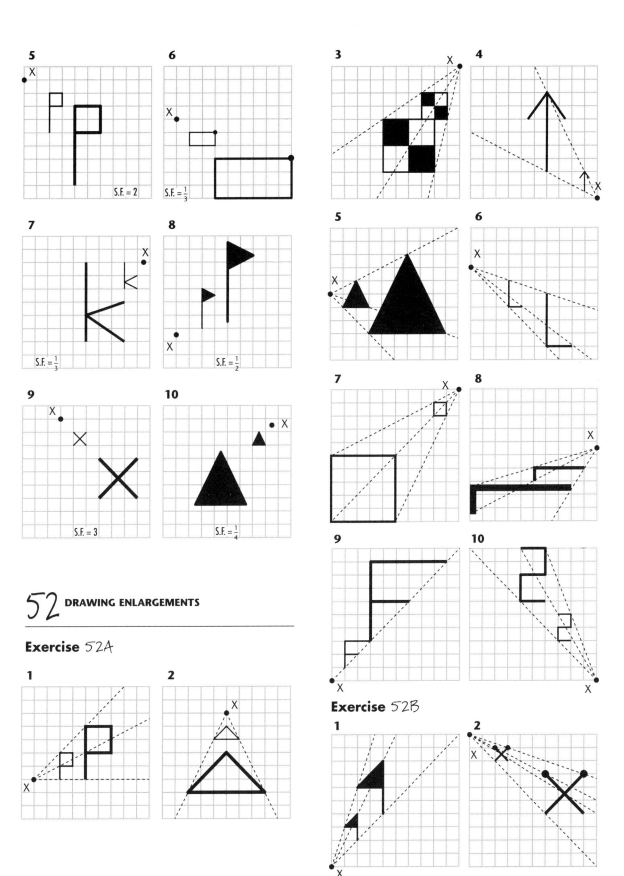

**5**

S.F. = 2

**6**

S.F. = $\frac{1}{3}$

**7**

S.F. = $\frac{1}{3}$

**8**

S.F. = $\frac{1}{2}$

**9**

S.F. = 3

**10**

S.F. = $\frac{1}{4}$

## 52 DRAWING ENLARGEMENTS

**Exercise** 52A

**1**

**2**

**3**

**4**

**5**

**6**

**7**

**8**

**9**

**10**

**Exercise** 52B

**1**

**2**

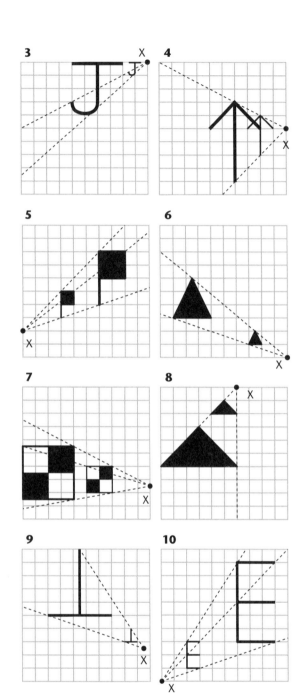

**3** **4** **5** **6** **7** **8** **9** **10**

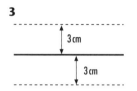
**3**
**4**

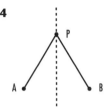
**5**
**6**

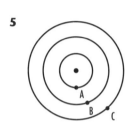
**7**
**8**

**9**

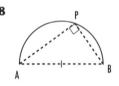

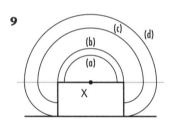

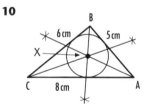
**10**

## 53 THE LOCUS OF A POINT

### Exercise 53A

**1**
**2**

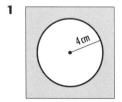

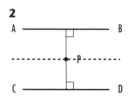

### Exercise 53B

**1** A perfect circle!

**2** A point equidistant from the 3 points, found by drawing a line from each point to the midpoint of a line joining the other 2 points.

**3** A straight line parallel to surface

**4**
**5**

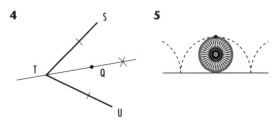

**6**

**7**

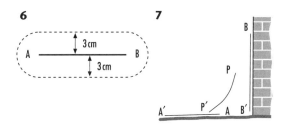

**8** A sphere

**9**

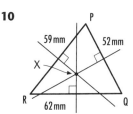

**10**

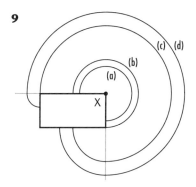

# 54 COMPOUND MEASURE

## Exercise 54A

| | | |
|---|---|---|
| **1** 37.13 g | **2** 62.5 km/h | **3** 170 m s$^{-1}$ |
| **4** 140 miles | **5** 136 miles | **6** 3.99 kg |
| **7** 120 km/h | **8** 2.07 g/cm$^3$ | **9** 250 miles |
| **10** 57.1 m.p.g. | **11** 1.41 cm$^3$ | **12** 7.1 g/cm$^3$ |
| **13** 35 m.p.h. | **14** 3 h | **15** 340 km |
| **16** 40 min | **17** 3 km/h | **18** 50 miles |
| **19** 11.9 km/$l$ | **20** 21.7 cm$^3$ | |

## Exercise 54B

| | | |
|---|---|---|
| **1** 26.25 g | **2** 71.4 m.p.h. | **3** 8.70 m s$^{-1}$ |
| **4** 66.3 g | **5** 3 h 20 min | **6** 2 h 44 min |
| **7** 30 km | **8** 47.1 m.p.g. | **9** 480 m s$^{-1}$ |
| **10** 5.0 cm$^3$ | **11** 50 min | **12** 290 km |
| **13** 9.07 km/$l$ | **14** 7.9 g/cm$^3$ | **15** 385 km |
| **16** 250 miles | **17** 19.3 g/cm$^3$ | **18** 21 miles |
| **19** 60 m.p.g. | **20** 32 cm$^3$ | |

# 55 NETS

## Exercise 55A

**1**

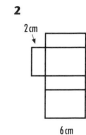

**2**

**3**

**4**

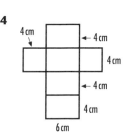

**5**

**6**

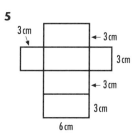

**7**

**8**

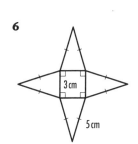

**9**

**10**

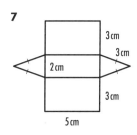

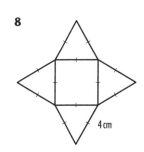

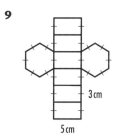

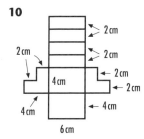

## Exercise 55B

**1**

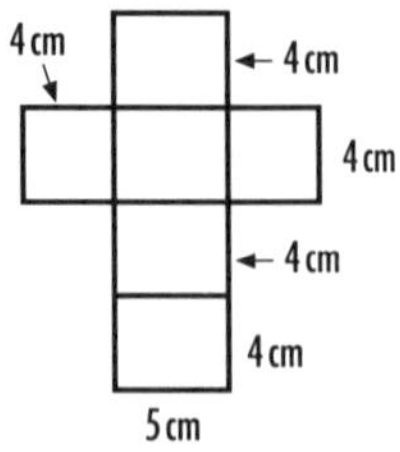


**2**

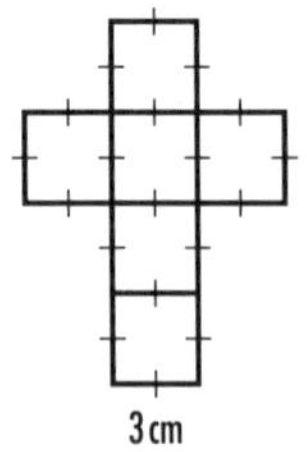


**3**

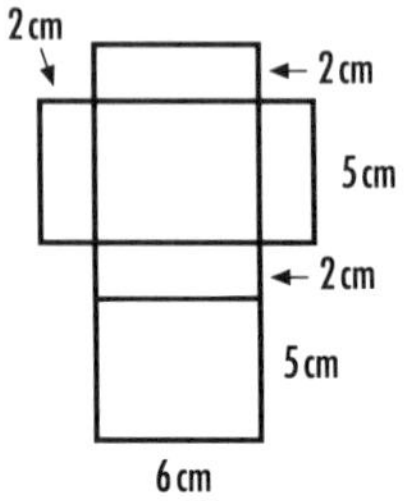


**4**

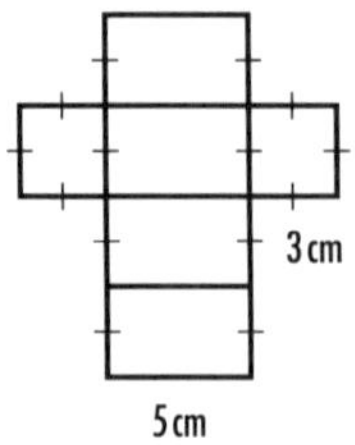


**5**

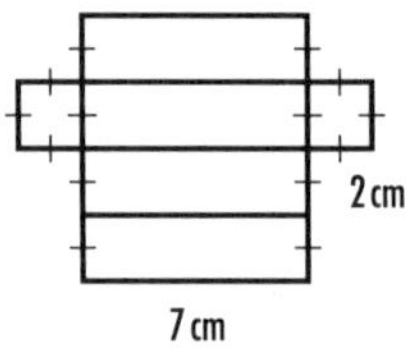


**6**

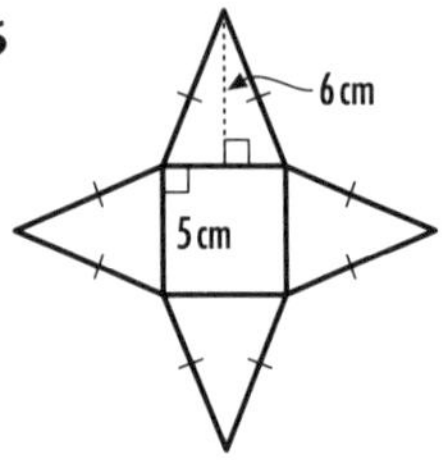


**7**

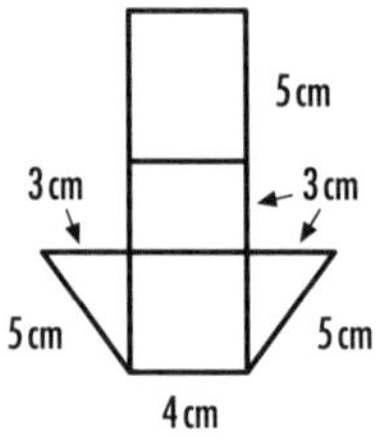


**8**

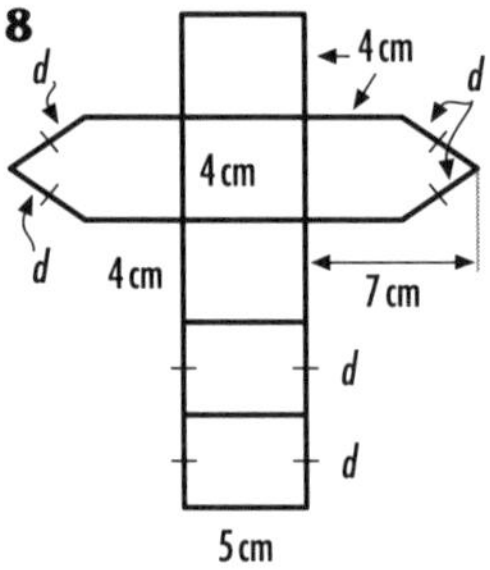


**9**

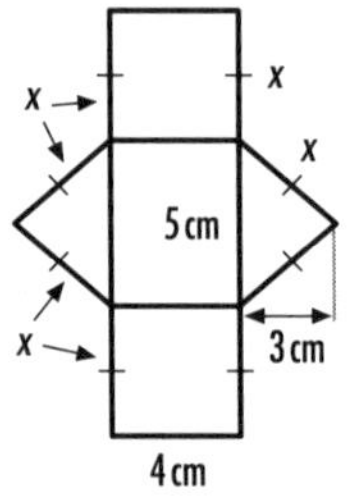


**10**

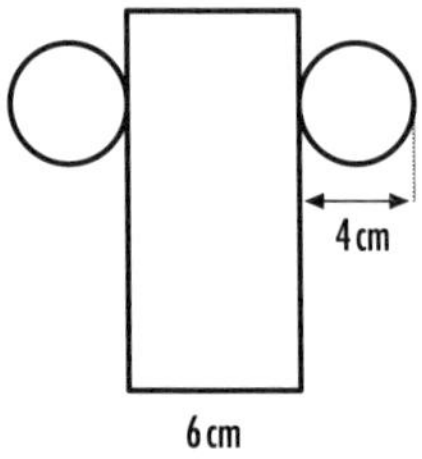


## Exercise 55C

| | | | |
|---|---|---|---|
| **1** E | **2** A | **3** B | **4** F |
| **5** D | **6** G | **7** C | |

## Exercise 55D

| | | | |
|---|---|---|---|
| **1** C | **2** D | **3** G | **4** E |
| **5** A | **6** B | **7** F | |

# 56 DRAWING SECTIONS THROUGH 3D SHAPES IN A GIVEN PLANE

## Exercise 56A

**1** (a)

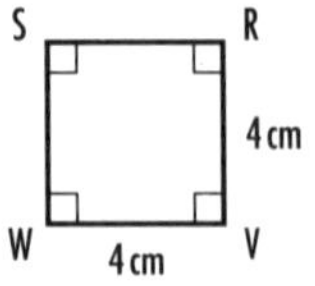


(b)

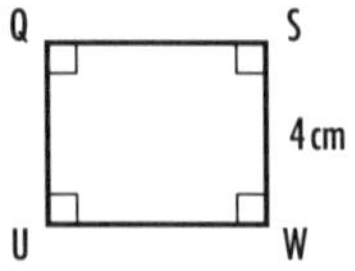


(c)

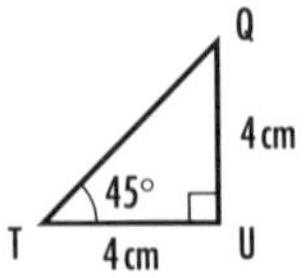


(d)

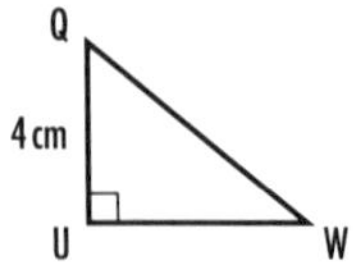


**2** (a)

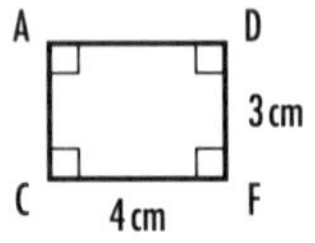


(b)

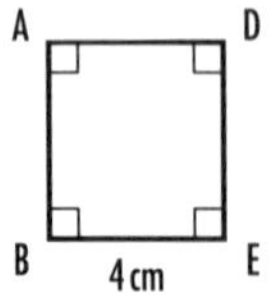


(c)

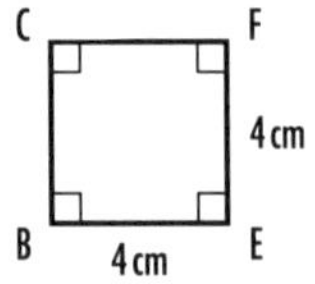


(d)

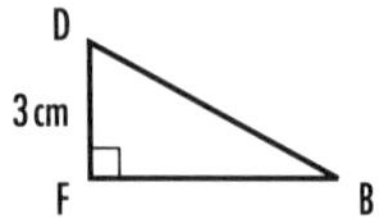


**3** (a)

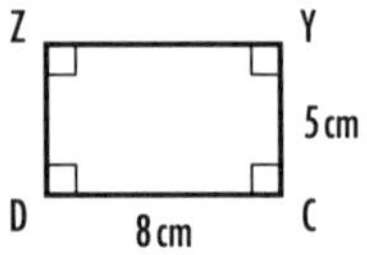


(b)

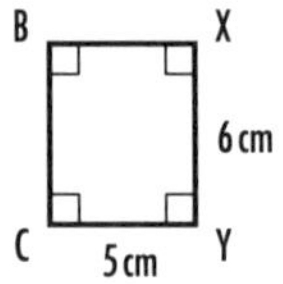


(c)

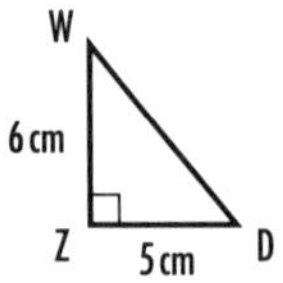


(d)

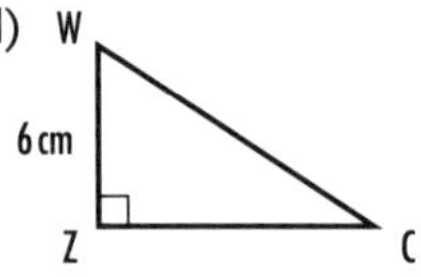


**4** (a)

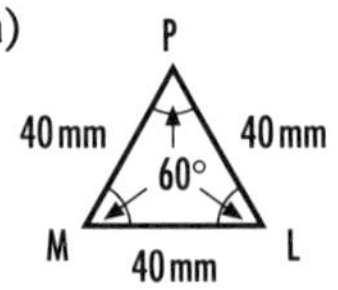


(b)

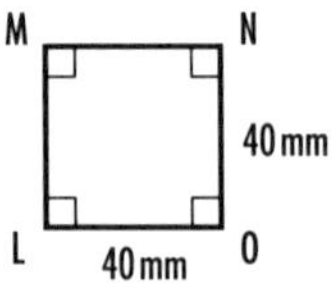


(c)

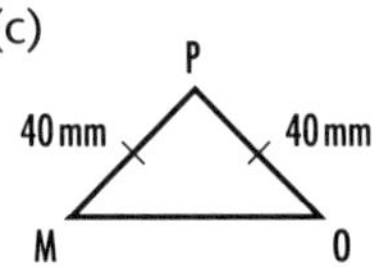


(d)

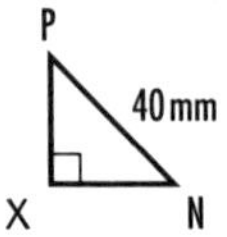

**5** (a)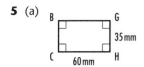

(b)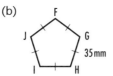

(c)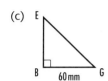

(d)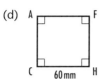

**6** (a)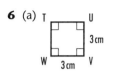

(b)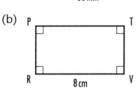

(c)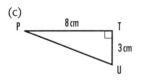

(d)

## Exercise 56B

**1** (a)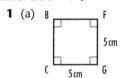

(b)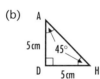

(c)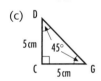

(d)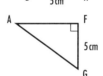

**2** (a)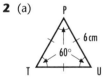

(b)

(c)

(d)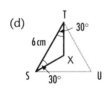

**3** (a)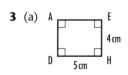

(b)

(c)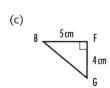

(d)

**4** (a)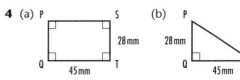

(c)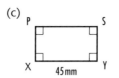

(d)

**5** (a)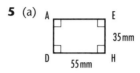

(b)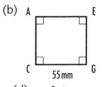

(c)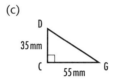

(d)

**6** (a)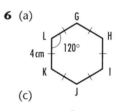

(b)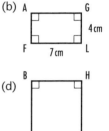

(c)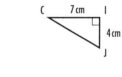

(d)

## REVISION

### Exercise E

**1** (a) $a = b = 70°$, $c = 110°$, $d = 110°$
(b) $e = 98°$, $f = 42°$, $g = 138°$
(c) $h = 50°$, $i = 58°$
(d) $j = 133°$
(e) $k = 51°$, $l = 105°$, $m = 54°$
(f) $n = p = q = 72°$, $r = 108°$

**2** (a) kite    (b) rhombus    (c) rectangle
(d) (regular) pentagon    (e) parallelogram
(f) isosceles trapezium    (g) (regular) hexagon

**3**

A B N X Y Z

**4** (a) 62 mm (b) 4 cm (c) 5 cm (d) 35 mm
**5** (a) 525 m (b) 17 km (c) 25 km (d) 154 km
**6** (a)

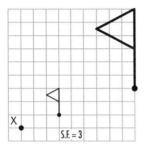

**(b)**

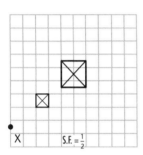

**7**

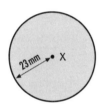

**8** 62.5 km/h
**9** 11.4 g/cm³
**10** 1 D; 2 B; 3 C; 4 A

**11** (a)

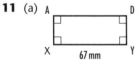

(b)

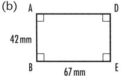

(c)

(d)

**Exercise EE**

**1** (a) $a = 36°$ (angles of a triangle),
   $b = 88°$ (corresponding angles),
   $c = 36°$ (corresponding angles)
   (b) $d = 30°$ (straight line),
   $e = 60°$ (vertically opposite angles),
   $f = 120°$ (straight line)

(c) $g = 60°$ (external angle of a hexagon),
   $h = 120°$ (straight line),
   $i = 60°$ (angles of a triangle)
(d) $j = 145°$ (alternate angles),
   $k = 35°$ (straight line),
   $l = 35°$ (base angles of isosceles trapezium)
(e) $m = 68°$ (corresponding angles),
   $n = 68°$ (vertically opposite),
   $p = 72°$ (supplementary angles),
   $q = 72°$ (straight line)
(f) $r = 90°$ (diagonals of rhombus),
   $s = 35°$ (base angles of isosceles triangle),
   $t = 35°$ (alternate, etc.),
   $u = 110°$ (supplementary, etc.)

**2**

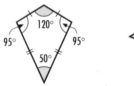

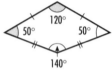

**3**

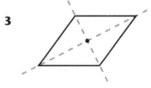

**4**

Pentagon, order = 5

**5** (a) 63 mm, 31.5 km (b) 45 mm, 4.5 km
**6** (a) 75 mm line (b) 13 mm line
**7** (a)

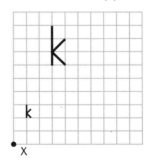

(b)

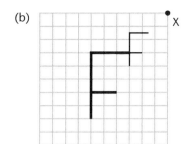

**8**

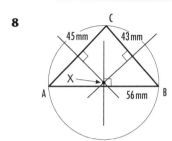

**9** 80.0 cm³

**10** (a)

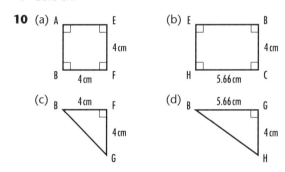

## Exercise 58A

| | | | | | |
|---|---|---|---|---|---|
| **1** | 9 cm | **2** | 24 m | **3** | 11.5 cm |
| **4** | 7.14 cm | **5** | 21 mm | **6** | 29.7 cm |
| **7** | 0.9 m | **8** | 1.52 m | **9** | 20.6 mm |
| **10** | 7.94 cm | **11** | 24 m | **12** | 30 mm |
| **13** | 0.825 m | **14** | 8.49 cm | **15** | 80 mm |
| **16** | 80.2 mm | **17** | 6.58 cm | **18** | 4.73 m |
| **19** | 5.91 m | **20** | 2.59 m | | |

## Exercise 58B

| | | | | | |
|---|---|---|---|---|---|
| **1** | 12 m | **2** | 16 cm | **3** | 3.79 m |
| **4** | 51.3 mm | **5** | 45 mm | **6** | 1.79 m |
| **7** | 84.9 m | **8** | 10.7 cm | **9** | 1.83 cm |
| **10** | 10 cm | **11** | 54.3 mm | **12** | 10.2 cm |
| **13** | 0.806 m | **14** | 13.2 cm | **15** | 2.90 m |
| **16** | 15.9 cm | **17** | 6.37 cm | **18** | 10.7 m |
| **19** | 41.6 mm | **20** | 109 mm | | |

## Exercise 59A

| | | | | | |
|---|---|---|---|---|---|
| **1** | 20 cm | **2** | 16 cm | **3** | 57.7 mm |
| **4** | 38.4 m | **5** | 130 m | **6** | 12 cm |
| **7** | 89.4 cm | **8** | 146 mm | **9** | 28.3 mm |
| **10** | 5.66 m | **11** | 10 cm | **12** | 5.29 m |
| **13** | 31.7 mm | **14** | 13 m | **15** | 4.95 m |
| **16** | 76.6 mm | **17** | 32.0 mm | **18** | 1.14 m |
| **19** | 7 cm | **20** | 60 mm | **21** | 7.62 cm |
| **22** | 22.4 mm | **23** | 13.3 cm | **24** | 2.41 m |
| **25** | 35 mm | **26** | 71.6 mm | **27** | 5.48 m |
| **28** | 3.12 cm | **29** | 3.38 m | **30** | 10 cm |

## Exercise 59B

| | | | | | |
|---|---|---|---|---|---|
| **1** | 25 cm | **2** | 14 mm | **3** | 105 mm |
| **4** | 4.91 m | **5** | 16 cm | **6** | 16.3 cm |
| **7** | 14.1 cm | **8** | 2.8 m | **9** | 156 cm |
| **10** | 142 mm | **11** | 43.0 mm | **12** | 5 cm |
| **13** | 15 m | **14** | 29.5 cm | **15** | 5.81 cm |
| **16** | 10.8 m | **17** | 45 mm | **18** | 1.72 m |
| **19** | 37.4 mm | **20** | 53.7 mm | **21** | 5.39 cm |
| **22** | 15.2 cm | **23** | 1.61 m | **24** | 7.07 cm |
| **25** | 99.2 mm | **26** | 80 mm | **27** | 0.6 m |
| **28** | 24.5 cm | **29** | 147 mm | **30** | 7.74 m |

## Exercise 57A

| | | | | | |
|---|---|---|---|---|---|
| **1** | 13 cm | **2** | 5 cm | **3** | 5.66 m |
| **4** | 54.1 mm | **5** | 29.2 mm | **6** | 10 cm |
| **7** | 17.7 cm | **8** | 2 m | **9** | 5.83 m |
| **10** | 106 mm | **11** | 8.65 cm | **12** | 17.8 m |
| **13** | 79.1 m | **14** | 8.08 cm | **15** | 14.1 cm |
| **16** | 114 mm | **17** | 25.5 cm | **18** | 4 m |
| **19** | 1.49 m | **20** | 42.2 mm | | |

## Exercise 57B

| | | | | | |
|---|---|---|---|---|---|
| **1** | 25 cm | **2** | 10 cm | **3** | 17.0 m |
| **4** | 7.21 m | **5** | 41.9 mm | **6** | 20 m |
| **7** | 22.4 cm | **8** | 29.2 m | **9** | 8 m |
| **10** | 1.13 m | **11** | 100 cm | **12** | 55.3 cm |
| **13** | 16.9 m | **14** | 2.47 m | **15** | 3.07 m |
| **16** | 2.83 m | **17** | 17.9 cm | **18** | 43.0 mm |
| **19** | 1.75 m | **20** | 5.30 m | | |

# 60 PROBLEMS INVOLVING PYTHAGORAS' THEOREM

## Exercise 60A

| | | |
|---|---|---|
| **1** 80 mm | **2** 60.8 cm | **3** 15 cm |
| **4** 66.7 mm | **5** 9.19 cm | **6** 5.47 m |
| **7** 5.66 units | **8** 96.0 mm | **9** 3.61 km |
| **10** 29.5 mm | **11** 8.04 m + 1 = 9.04 m | |
| **12** 22.7 m | **13** 3.12 cm, 5.45 cm | |
| **14** 2.83 units | **15** 4.83 cm | |

## Exercise 60B

| | | |
|---|---|---|
| **1** 11.3 cm | **2** 13.4 cm | **3** 6.32 m |
| **4** 25 cm | **5** 3.5 m | **6** 18.0 cm |
| **7** 2.26 m | **8** 9.97 cm | **9** 358 mm |
| **10** 8 cm | | |
| **11** CD = 1.41 units, AE = 2.24 units | | |
| **12** 49.8 cm | **13** $x$ = 15 cm, $y$ = 11.2 cm | |
| **14** 8 cm | **15** 7.21 cm | |

# 61 AREA: QUADRILATERALS AND TRIANGLES

## Exercise 61A

| | |
|---|---|
| **1** 56 cm$^2$ | **2** 13 000 mm$^2$ |
| **3** 60 cm$^2$ | **4** 15.8 cm$^2$ |
| **5** 6.25 cm$^2$ | **6** 187 cm$^2$ |
| **7** 0.68 m$^2$ or 6800 cm$^2$ | |
| **8** 87 500 mm$^2$ or 8.75 m$^2$ | |
| **9** 540 mm$^2$ or 5.4 cm$^2$ | |
| **10** 1344 mm$^2$ or 13.44 cm$^2$ | |
| **11** 56 cm$^2$ | |
| **12** 6144 mm$^2$ or 61.44 cm$^2$ | |
| **13** 24 cm$^2$ | **14** 90 cm$^2$ |
| **15** 682 mm$^2$ or 6.82 cm$^2$ | **16** 45 cm$^2$ |
| **17** 10 cm$^2$ | **18** 1.69 m$^2$ |
| **19** 16 cm$^2$ | **20** 1512 mm$^2$ |
| **21** 16 cm | **22** 6.52 cm |
| **23** 15 cm | **24** 1.8 m or 180 cm |
| **25** 29.3 mm | **26** 50 mm or 5 cm |
| **27** 16 cm | **28** 9.22 cm |
| **29** 12.5 cm | **30** 3.75 cm or 37.5 mm |

## Exercise 61B

| | |
|---|---|
| **1** 120 cm$^2$ | **2** 1620 mm$^2$ |
| **3** 8925 mm$^2$ | **4** 51.84 cm$^2$ |
| **5** 0.875 cm$^2$ | |
| **6** 14850 mm$^2$ or 148.5 cm$^2$ | |
| **7** 65 cm$^2$ | |
| **8** 4950 mm$^2$ or 49.5 cm$^2$ | |
| **9** 4800 mm$^2$ or 48 cm$^2$ | |

| | | |
|---|---|---|
| **10** 30.24 cm$^2$ or 3024 mm$^2$ | | |
| **11** 30 cm$^2$ | **12** 0.64 m$^2$ | |
| **13** 108 cm$^2$ | **14** 25.2 cm$^2$ | |
| **15** 1080 mm$^2$ or 10.8 cm$^2$ | **16** 2.09 m$^2$ | |
| **17** 18 cm$^2$ | **18** 7.5 m$^2$ | |
| **19** 378 mm$^2$ | **20** 24 cm$^2$ | |
| **21** 12.2 cm | **22** 2.4 cm | |
| **23** 30.00 m or 3 cm | **24** 3.61 cm | |
| **25** 1.40 m | **26** 30 mm | |
| **27** 8 cm | **28** 80 mm or 8 cm | |
| **29** 267 cm | **30** 0.5 cm | |

# 62 CIRCUMFERENCE OF A CIRCLE

## Exercise 62A

| | | |
|---|---|---|
| **1** 37.7 cm | **2** 132 mm | **3** 5.03 m |
| **4** 101 cm | **5** 138 mm | **6** 81.7 cm |
| **7** 8.17 m | **8** 34.6 cm | **9** 408 mm |
| **10** 88.0 cm | **11** 47.1 cm | **12** 23.6 cm |
| **13** 302 mm | **14** 3.77 m | **15** 81.7 cm |
| **16** 339 mm | **17** 170 mm | **18** 4.40 m |
| **19** 28.3 cm | **20** 73.5 cm | **21** 176 mm |
| **22** 9.43 m | **23** 14.1 cm | **24** 20.1 m |
| **25** 25.1 cm | **26** 66.0 cm | **27** 157 cm |
| **28** 408 cm | **29** C = 40.8 m, 5 packs | |
| **30** 7.86 m | | |

## Exercise 62B

| | | |
|---|---|---|
| **1** 31.4 cm | **2** 18.8 cm | **3** 88.0 mm |
| **4** 41.5 m | **5** 18.8 cm | **6** 22.0 cm |
| **7** 5.65 m | **8** 195 mm | **9** 298 mm |
| **10** 44.0 cm | **11** 50.3 cm | **12** 4.08 m |
| **13** 415 mm | **14** 17.6 cm | **15** 33.0 cm |
| **16** 13.2 m | **17** 339 mm | **18** 245 mm |
| **19** 50.3 cm | **20** 22.6 m | **21** 50.3 cm |
| **22** 17.0 m | **23** 53.4 cm | **24** 59.7 cm |
| **25** 101 cm | **26** 19.5 m | **27** 141 mm |
| **28** 377 mm | **29** 63 cm | **30** 12.6 m |

# 63 AREA OF A CIRCLE

## Exercise 63A

| | | |
|---|---|---|
| **1** 113 cm$^2$ | **2** 1390 mm$^2$ | **3** 2.01 m$^2$ |
| **4** 804 cm$^2$ | **5** 1520 mm$^2$ | **6** 531 cm$^2$ |
| **7** 5.31 m$^2$ | **8** 95.0 cm$^2$ | **9** 13 300 mm$^2$ |
| **10** 616 cm$^2$ | **11** 177 cm$^2$ | **12** 44.2 cm$^2$ |
| **13** 7240 mm$^2$ | **14** 1.13 m$^2$ | **15** 531 cm$^2$ |
| **16** 9160 mm$^2$ | **17** 2290 mm$^2$ | **18** 1.54 m$^2$ |
| **19** 63.6 cm$^2$ | **20** 430 cm$^2$ | **21** 2460 mm$^2$ |

| | | |
|---|---|---|
| **22** $7.07\,m^2$ | **23** $15.9\,cm^2$ | **24** $32.2\,m^2$ |
| **25** $314\,cm^2$ | **26** $177\,cm^2$ | **27** $28.3\,m^2$ |
| **28** $3850\,cm^2$ | **29** $2.26\,m^2$ | **30** $80\,kg$ |

## Exercise 63B

| | | |
|---|---|---|
| **1** $78.5\,cm^2$ | **2** $28.3\,cm^2$ | **3** $616\,mm^2$ |
| **4** $137\,m^2$ | **5** $28.3\,cm^2$ | **6** $38.5\,cm^2$ |
| **7** $2.54\,m^2$ | **8** $3020\,mm^2$ | **9** $7090\,mm^2$ |
| **10** $154\,cm^2$ | **11** $201\,cm^2$ | **12** $1.33\,m^2$ |
| **13** $13\,700\,mm^2$ | **14** $24.6\,cm^2$ | **15** $86.6\,cm^2$ |
| **16** $13.9\,m^2$ | **17** $9160\,mm^2$ | **18** $4780\,mm^2$ |
| **19** $201\,cm^2$ | **20** $40.7\,m^2$ | **21** $201\,cm^2$ |
| **22** $22.9\,m^2$ | **23** $227\,cm^2$ | **24** $284\,cm^2$ |
| **25** $255\,cm^2$ | **26** $71\,m^2$ | **27** $227\,m^2$ |
| **28** $1.13\,m^2$ | **29** $491\,m^2$ | **30** $2640\,mm^2$ |

# 64 VOLUME: CUBE AND CUBOIDS

## Exercise 64A

| | |
|---|---|
| **1** $125\,cm^3$ | **2** $432\,mm^3$ |
| **3** $42\,cm^3$ | **4** $216\,cm^3$ |
| **5** $7680\,mm^3$ | **6** $0.24\,m^3$ |
| **7** $91\,125\,mm^3$ | **8** $308\,cm^3$ |
| **9** $24.6\,cm^3$ | **10** $0.729\,m^3$ |
| **11** $96\,000\,mm^3$ | **12** $27.6\,cm^3$ |
| **13** $21\,840\,cm^3$ | **14** $1.14\,m^3$ |
| **15** $275\,cm^3$ | **16** $116\,550\,mm^3$ |
| **17** $13\,800\,cm^3$ | **18** $2744\,mm^3$ |
| **19** $111\,cm^3$ | **20** $1728\,cm^3$ |

## Exercise 64B

| | |
|---|---|
| **1** $160\,cm^3$ | **2** $216\,cm^3$ |
| **3** $3240\,mm^3$ | **4** $0.432\,m^3$ |
| **5** $422\,m^3$ | **6** $24.5\,cm^3$ |
| **7** $48.5\,m^3$ | **8** $26\,880\,mm^3$ |
| **9** $0.125\,m^3$ | **10** $20\,000\,cm^3$ |
| **11** $6.48\,cm^3$ | **12** $3375\,mm^3$ |
| **13** $168\,cm^3$ | **14** $0.08\,m^3$ |
| **15** $26\,250\,cm^3$ | **16** $300\,763\,mm^3$ |
| **17** $16\,200\,cm^3$ | **18** $56.9\,m^3$ |
| **19** $22.0\,cm^3$ | **20** $437\,cm^3$ |

## Exercise 64C

| | | |
|---|---|---|
| **1** $7\,cm$ | **2** $6\,mm$ | **3** $7\,cm$ |
| **4** $6.5\,cm$ | **5** $10\,cm$ | **6** $4.64\,cm$ |
| **7** $7\,cm$ | **8** $0.5\,m$ | **9** $9\,cm$ |
| **10** $30\,mm$ | **11** $2\,m$ | **12** $5\,cm$ |
| **13** (a) $18\,cm^3$ | (b) $18\,000\,mm^3$ | |
| **14** $30\,cm$ | **15** $8\,cm$ | |
| **16** (a) $3.375\,m^3$ | (b) $3\,375\,000\,cm^3$ | (c) $3375\,l$ |
| **17** $2.5\,m$ | | |

| | | |
|---|---|---|
| **18** (a) $5.94\,m^3$ | (b) $5\,940\,000\,cm^3$ | (c) $5940\,l$ |
| **19** $70\,cm$ | **20** $19\,cm$ | |

## Exercise 64D

| | | |
|---|---|---|
| **1** $7.07\,cm$ | **2** $7\,cm$ | **3** $32\,mm$ |
| **4** $0.4\,m$ | **5** $5\,cm$ | **6** $8\,cm$ |
| **7** $0.75\,m$ | **8** $15\,cm$ | **9** $12\,cm$ |
| **10** $8.5\,cm$ | **11** $3.0\,cm$ | |
| **12** (a) $816\,cm^3$ | (b) $816\,000\,mm^3$ | |
| **13** $10\,cm$ | | |
| **14** (a) $238\,000\,cm^3$ | (b) $238\,000\,000\,mm^3$ | |
| (c) $238\,l$ | | |
| **15** $12.6\,cm$ | **16** $1.2\,m$ | **17** $15\,cm$ |
| **18** (a) $4.096\,m^3$ | (b) $4\,096\,000\,cm^3$ | (c) $4096\,l$ |
| **19** $45\,mm$ | | |
| **20** (a) $798\,000\,cm^3$ | (b) $0.798\,m^3$ | (c) $798\,l$ |

# 65 VOLUME OF A PRISM USING AREA OF CROSS-SECTION

## Exercise 65A

| | | |
|---|---|---|
| **1** $225\,cm^3$ | **2** $192\,cm^3$ | **3** $51\,cm^3$ |
| **4** $1\,m^3$ | **5** $3625\,mm^3$ | **6** $11.34\,cm^3$ |
| **7** $1.43\,m^3$ | **8** $712.5\,cm^3$ | **9** $421.6\,cm^3$ |
| **10** $72\,000\,mm^3$ | **11** $283.5\,cm^3$ | **12** $0.725\,m^3$ |
| **13** $324\,cm^3$ | **14** $6720\,mm^3$ | **15** $43.2\,cm^3$ |
| **16** $56\,000\,mm^3$ | **17** $279\,cm^3$ | **18** $180\,cm^3$ |
| **19** $1.04\,m^3$ | **20** $252\,cm^3$ | |

## Exercise 65B

| | | |
|---|---|---|
| **1** $60\,cm^3$ | **2** $504\,cm^3$ | **3** $3480\,mm^3$ |
| **4** $726\,cm^3$ | **5** $2.31\,m^3$ | **6** $0.045\,m^3$ |
| **7** $1449\,cm^3$ | **8** $14\,400\,mm^3$ | **9** $72.8\,cm^3$ |
| **10** $0.76\,m^3$ | **11** $98.8\,cm^3$ | **12** $75\,900\,mm^3$ |
| **13** $350\,cm^3$ | **14** $4250\,mm^3$ | **15** $1.2\,m^3$ |
| **16** $216.2\,cm^3$ | **17** $12\,600\,mm^3$ | **18** $0.64\,m^3$ |
| **19** $88.4\,cm^3$ | **20** $21\,840\,mm^3$ | |

## Exercise 65C

| | | |
|---|---|---|
| **1** $20\,cm$ | **2** $0.11\,m^2$ | **3** $25\,cm^2$ |
| **4** $16\,mm$ | **5** $50\,cm^2$ | **6** $0.35\,m$ |
| **7** $8\,cm$ | **8** $80\,mm^2$ | **9** $17.5\,cm^2$ |
| **10** $2\,m^2$ | **11** $1.5\,cm$ | **12** $250\,mm^2$ |
| **13** $36\,cm^2$ | **14** $8\,cm$ | **15** $0.25\,m$ |
| **16** $4.4\,m^2$ | **17** $1.25\,cm$ | **18** $6\,mm$ |
| **19** $400\,mm$ | **20** $0.128\,m^2$ | |

## Exercise 65D

| | | |
|---|---|---|
| **1** $21\,cm^2$ | **2** $1.75\,cm$ | **3** $0.9\,mm$ |
| **4** $3.2\,m^2$ | **5** $18\,cm^2$ | **6** $9\,mm$ |
| **7** $0.9\,cm^2$ | **8** $160\,m$ | **9** $11.5\,cm^2$ |
| **10** $6\,cm$ | **11** $15\,mm$ | **12** $1.2\,m^2$ |

**13** $12 \, \text{cm}^2$ **14** $5 \, \text{cm}$ **15** $2.5 \, \text{cm}$

**16** $20 \, \text{mm}^2$ **17** $70 \, \text{mm}$

**18** $2.12 \, \text{m}$ (3 s.f.) **19** $0.8 \, \text{m}^2$ **20** $50 \, \text{mm}^2$

# 66 VOLUME OF A PRISM, INCLUDING THE CYLINDER

## Exercise 66A

**1** $108 \, \text{cm}^3$ **2** $64.8 \, \text{cm}^3$ **3** $2260 \, \text{cm}^3$

**4** $540 \, \text{cm}^3$ **5** $396 \, \text{cm}^3$ **6** $185 \, \text{cm}^3$

**7** $288 \, \text{cm}^3$ **8** $1680 \, \text{cm}^3$ **9** $56.5 \, \text{cm}^3$

**10** $920 \, \text{cm}^3$ **11** $30 \, \text{cm}^3$ **12** $1560 \, \text{cm}^3$

**13** $360 \, \text{cm}^3$ **14** $581 \, \text{cm}^3$ **15** $336 \, \text{cm}^3$

**16** $5010 \, \text{cm}^3$ **17** $11.0 \, \text{cm}$ **18** $7 \, \text{cm}$

**19** $172 \, \text{cm}^3$ **20** (a) $24 \, \text{cm}^2$ (b) $288 \, \text{cm}^3$

## Exercise 66B

**1** $160 \, \text{cm}^3$ **2** $112 \, \text{cm}^3$ **3** $943 \, \text{cm}^3$

**4** $189 \, \text{cm}^3$ **5** $252 \, \text{cm}^3$ **6** $216 \, \text{cm}^3$

**7** $176 \, \text{cm}^3$ **8** $1120 \, \text{cm}^3$ **9** $1440 \, \text{cm}^3$

**10** $3460 \, \text{cm}^3$ **11** $216 \, \text{cm}^3$ **12** $503 \, \text{cm}^3$

**13** $616 \, \text{cm}^2$ **14** $15.8 \, \text{cm}^3$ **15** $7 \, \text{cm}$

**16** $882 \, \text{cm}^3$ **17** $296 \, \text{cm}^3$ **18** $9.6 \, \text{cm}$

**19** $0.942 \, \text{m}^3$ **20** $252 \, 000 \, \text{mm}^3$ (b) $252 \, \text{cm}^3$

# 67 SURFACE AREAS OF SHAPES

## Exercise 67A

**1** $150 \, \text{cm}^2$ **2** $158 \, \text{cm}^2$ **3** $252 \, \text{cm}^2$

**4** $168 \, \text{cm}^2$ **5** $348 \, \text{cm}^2$ **6** $8160 \, \text{mm}^2$

**7** $162 \, \text{cm}^2$ **8** $972 \, \text{mm}^2$ **9** $40 \, \text{cm}^2$

**10** $30.5 \, \text{m}^2$ **11** $127.8 \, \text{cm}^2$ **12** $528 \, \text{cm}^2$

## Exercise 67B

**1** $232 \, \text{cm}^2$ **2** $193.2 \, \text{cm}^2$ **3** $121.5 \, \text{cm}^2$

**4** $118 \, \text{cm}^2$ **5** $328 \, \text{cm}^2$ **6** $182 \, \text{cm}^2$

**7** $17 \, 000 \, \text{mm}^2$ or $170 \, \text{cm}^2$ **8** $736 \, \text{cm}^2$

**9** $79.68 \, \text{cm}^2$ or $7968 \, \text{mm}^2$ **10** $408 \, \text{cm}^2$

**11** $45.6 \, \text{m}^2$ **12** $270 \, \text{cm}^2$

# REVISION

## Exercise F

**1** (a) $10 \, \text{cm}$ (b) $5 \, \text{cm}$ (c) $20 \, \text{cm}$

(d) $25 \, \text{cm}$

**2** (a) $120 \, \text{cm}^2$ (b) $56 \, \text{cm}^2$ (c) $42 \, \text{cm}^2$

(d) $85 \, \text{cm}^2$ (e) $1024 \, \text{mm}^2$

**3** (a) $37.7 \, \text{cm}$ (b) $154 \, \text{cm}^2$ (c) $78.5 \, \text{cm}^2$

(d) $94.2 \, \text{mm}$

**4** (a) $125 \, \text{cm}^2$ (b) $60 \, \text{cm}^2$

**5** (a) $60 \, \text{cm}^3$, $104 \, \text{cm}^2$ (b) $75.6 \, \text{cm}^3$, $127 \, \text{cm}^2$

(c) $200 \, \text{cm}^3$, $240 \, \text{cm}^2$

**6** $226 \, \text{cm}^3$

## Exercise FF

**1** (a) $10 \, \text{cm}$ (b) $11.2 \, \text{cm}$

**2** $84.9 \, \text{mm}$

**3** (a) $2.94 \, \text{m}$

(b) $2.94 - 2.68 = 0.26 \, \text{m}$ or $0.258 \, \text{m}$ (to 3 s.f.)

**4** $9.4 \, \text{cm}$

**5** (a) $1090 \, \text{cm}^2$ (b) $60.7\%$

**6** $0.637 \, \text{m}$

**7** (a) $109 \, 200 \, \text{cm}^3$ (b) $0.1092 \, \text{m}^3$ (c) $109.2$ litres

**8** (a) $160 \, \text{m}^2$ (b) $2250 \, \text{mm}^2$

**9** $61.4 \, \text{m}^2$

**10** (a) $13 \, \text{cm}$

(b) sketches; $13 \, \text{cm}$ as hypotenuse, $13 \, \text{cm}$ as ordinary side

(c) $30 \, \text{cm}^2$ and $32.5 \, \text{cm}^2$

(d) $360 \, \text{cm}^3$ and $390 \, \text{cm}^3$

(e) $420 \, \text{cm}^2$ and $448 \, \text{cm}^2$

# Handling data

**68** CONTINUOUS DATA: CREATING FREQUENCY
TABLES AND FREQUENCY DIAGRAMS

## Exercise 68A

**1**

| Age, A (years) | Frequency |
|---|---|
| $10 \leq A < 20$ | 7 |
| $20 \leq A < 30$ | 10 |
| $30 \leq A < 40$ | 6 |
| $40 \leq A < 50$ | 4 |
| $50 \leq A < 60$ | 3 |
| $60 \leq A < 70$ | 2 |
| $70 \leq A < 80$ | 2 |

**2**

| Speed, S (m.p.h.) | Frequency |
|---|---|
| $10 \leq S < 20$ | 2 |
| $20 \leq S < 30$ | 8 |
| $30 \leq S < 40$ | 19 |
| $40 \leq S < 50$ | 7 |
| $50 \leq S < 60$ | 3 |
| $60 \leq S < 70$ | 1 |

**3**

| Length, L (cm) | Frequency |
|---|---|
| $25 \leq L < 30$ | 1 |
| $30 \leq L < 35$ | 4 |
| $35 \leq L < 40$ | 6 |
| $40 \leq L < 45$ | 10 |
| $45 \leq L < 50$ | 8 |
| $50 \leq L < 55$ | 3 |

**4**

| Volume, v (ml) | Frequency |
|---|---|
| $0 \leq v < 10$ | 10 |
| $10 \leq v < 20$ | 13 |
| $20 \leq v < 30$ | 8 |
| $30 \leq v < 40$ | 4 |
| $40 \leq v < 50$ | 3 |
| $50 \leq v < 60$ | 2 |

**5**

| Distance, d (m) | Frequency |
|---|---|
| $35 \leq d < 40$ | 2 |
| $40 \leq d < 45$ | 4 |
| $45 \leq d < 50$ | 7 |
| $50 \leq d < 55$ | 7 |
| $55 \leq d < 60$ | 13 |
| $60 \leq d < 65$ | 2 |

**6**

| Height, h (m) | Frequency |
|---|---|
| $0.50 \leq h < 1.00$ | 3 |
| $1.00 \leq h < 1.50$ | 5 |
| $1.50 \leq h < 2.00$ | 2 |
| $2.00 \leq h < 2.50$ | 21 |

**7**

| Mass, m (gram) | Frequency |
|---|---|
| $39.6 \leq m < 39.7$ | 1 |
| $39.7 \leq m < 39.8$ | 2 |
| $39.8 \leq m < 39.9$ | 5 |
| $39.9 \leq m < 40.0$ | 7 |
| $40.0 \leq m < 40.1$ | 11 |
| $40.1 \leq m < 40.2$ | 7 |
| $40.2 \leq m < 40.3$ | 2 |
| $40.3 \leq m < 40.4$ | 1 |

**8**

| Area, A (m²) | Frequency |
|---|---|
| $0 \leq A < 5$ | 7 |
| $5 \leq A < 10$ | 24 |
| $10 \leq A < 15$ | 11 |
| $15 \leq A < 20$ | 4 |
| $20 \leq A < 25$ | 2 |
| $25 \leq A < 30$ | 2 |

**9** Answers vary

**10** Answers vary

## Exercise 68B

**1**

| Length, L (mm) | Frequency |
|---|---|
| $35 \leq L < 40$ | 4 |
| $40 \leq L < 45$ | 6 |
| $45 \leq L < 50$ | 18 |
| $50 \leq L < 55$ | 8 |
| $55 \leq L < 60$ | 3 |
| $60 \leq L < 65$ | 1 |

**2**

| Mass, $m$ (g) | Frequency |
|---|---|
| $40 \leq m < 50$ | 3 |
| $50 \leq m < 60$ | 7 |
| $60 \leq m < 70$ | 10 |
| $70 \leq m < 80$ | 5 |
| $80 \leq m < 90$ | 3 |
| $90 \leq m < 100$ | 3 |
| $100 \leq m < 110$ | 2 |
| $110 \leq m < 120$ | 2 |

**3**

| Rainfall, $R$ (mm) | Frequency |
|---|---|
| $0 \leq R < 20$ | 6 |
| $20 \leq R < 40$ | 10 |
| $40 \leq R < 60$ | 10 |
| $60 \leq R < 80$ | 5 |
| $80 \leq R < 100$ | 3 |
| $100 \leq R < 120$ | 2 |

**4**

| Height, $H$ (cm) | Frequency |
|---|---|
| $130 \leq H < 140$ | 2 |
| $140 \leq H < 150$ | 5 |
| $150 \leq H < 160$ | 8 |
| $160 \leq H < 170$ | 12 |
| $170 \leq H < 180$ | 3 |
| $180 \leq H < 190$ | 2 |

**5**

| Yield, $Y$ (kg) | Frequency |
|---|---|
| $1.0 \leq Y < 1.5$ | 3 |
| $1.5 \leq Y < 2.0$ | 3 |
| $2.0 \leq Y < 2.5$ | 6 |
| $2.5 \leq Y < 3.0$ | 9 |
| $3.0 \leq Y < 3.5$ | 8 |
| $3.5 \leq Y < 4.0$ | 7 |
| $4.0 \leq Y < 4.5$ | 3 |
| $4.5 \leq Y < 5.0$ | 1 |

**6**

| Handspan, $s$ (cm) | Frequency |
|---|---|
| $12 \leq s < 14$ | 3 |
| $14 \leq s < 16$ | 5 |
| $16 \leq s < 18$ | 13 |
| $18 \leq s < 20$ | 7 |
| $20 \leq s < 22$ | 3 |
| $22 \leq s < 24$ | 2 |

**7**

| Time, $T$ (s) | Frequency |
|---|---|
| $10.5 \leq T < 10.6$ | 2 |
| $10.6 \leq T < 10.7$ | 5 |
| $10.7 \leq T < 10.8$ | 12 |
| $10.8 \leq T < 10.9$ | 6 |
| $10.9 \leq T < 11.0$ | 2 |
| $11.0 \leq T < 11.1$ | 1 |
| $11.1 \leq T < 11.2$ | 1 |
| $11.2 \leq T < 11.3$ | 1 |
| $11.3 \leq T < 11.4$ | 1 |
| $11.4 \leq T < 11.5$ | 1 |

**8**

| Temperature, $T$ (°C) | Frequency |
|---|---|
| $15 \leq T < 20$ | 5 |
| $20 \leq T < 25$ | 7 |
| $25 \leq T < 30$ | 25 |
| $30 \leq T < 35$ | 18 |
| $35 \leq T < 40$ | 7 |

**9** Answers vary
**10** Answers vary

## 69 CONTINUOUS DATA: INTERPRETING FREQUENCY DIAGRAMS

### Exercise 69A
**1** (a) 12 hours    (b) 37    (c) 16.1 %
   (d) 109 hours
**2** (a) 28    (b) 5 kg    (c) 16    (d) 28.6%
**3** (a) 40    (b) 2    (c) 5    (d) 20%
**4** (a) June    (b) 2°C    (c) 18 days   (d) 3.28%
**5** (a) 60    (b) 24    (c) 17    (d) 40%
**6** (a) 87    (b) 10    (c) 30    (d) 21.8%

### Exercise 69B
**1** (a) 52    (b) 8    (c) 19    (d) 13.5%
**2** (a) 5 days    (b) $8 \leq h < 10$ hours
   (c) 22 days    (d) $\frac{1}{3}$
**3** (a) 120    (b) 2 kg    (c) 24    (d) 41.7%
**4** (a) 140    (b) 32    (c) 16    (d) $\frac{1}{5}$
**5** (a) 165    (b) 7    (c) 60    (d) 63.6%
**6** (a) 60    (b) 10 cm    (c) 18    (d) 38.3%

# 70 FREQUENCY POLYGONS: CONSTRUCTION

## Exercise 70A

**1**

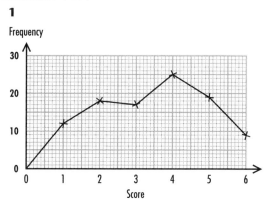

**2**

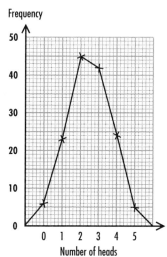

**3**

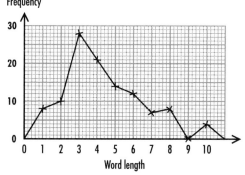

**4**

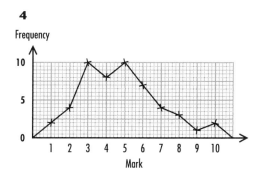

**5**

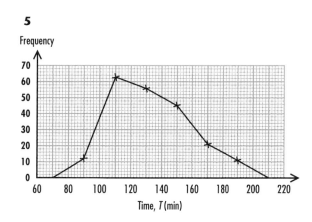

**6**

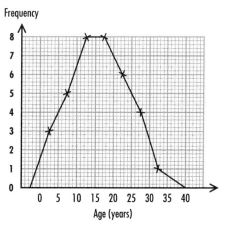

**7**

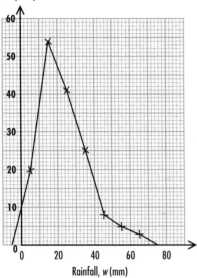

**3**

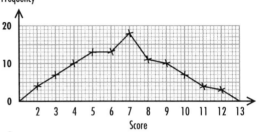

**4**

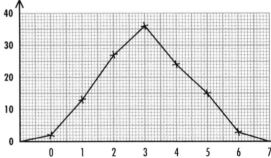

**Exercise** 70B

**1**

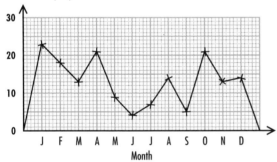

**5**

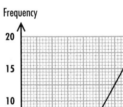

**2**

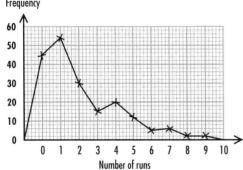

**6**

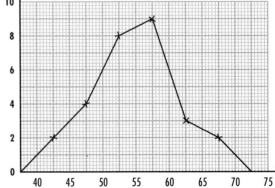

**7**

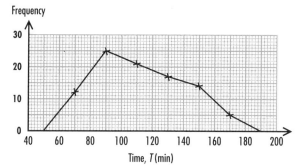

Frequency

Time, *T* (min)

**71 INTERPRETING AND COMPARING FREQUENCY POLYGONS**

## Exercise 71A

**1** (a) True
  (b) True
  (c) True
  (d) False – 3 heads = 2 tails and Sam's coins were below the theoretical value at this point.

**2** (a) True
  (b) False – both A and B contain 120 words.
  (c) True
  (d) True

**3** (a) False – Batch 1 has 100 plants and Batch 2 has 90 plants.
  (b) False
  (c) False – there are only 7 with heights in the class 85 ≤ h < 90 cm.
  (d) True

**4** (a) False – the population of the UK town is approximately double the size of the Kenyan town.
  (b) False – Kenyan town has smaller *number* in this age group but the total population is only half that of the UK town.
  (c) True
  (d) False – the lines have no meaning except on the plotted points.

**5** (a) True
  (b) False – the tests are the same, the people change.
  (c) False – the lines have no meaning except on the plotted points.
  (d) True

**6** (a) False – 5 less of brand B
  (b) True
  (c) True
  (d) False – 28 broke within the class interval 0.89 to 0.90 kg.

## Exercise 71B

**1** (a) True
  (b) True
  (c) False – dice B appears biased towards 4.
  (d) True

**2** (a) True
  (b) True
  (c) False – the monthly average is always above 0°C.
  (d) False – January has a far greater difference.

**3** (a) False – the graph shows how many weeks, not which weeks.
  (b) False – there are more weeks with good sales in 1995.
  (c) False – £250 is only used as the midpoint of £0 ≤ Sales < £500
  (d) True

**4** (a) True
  (b) False – they both drive for 125 days.
  (c) False – it shows they both have 8 days in which they drive between 250 to 300 miles.
  (d) True

**5** (a) True
  (b) True
  (c) False – the lines have no meaning except at the plotted points.
  (d) True

**6** (a) False – there are more samples from machine A.
  (b) False – shows there are equal numbers in the class interval.
  (c) False – over 30% cannot be sold.
  (d) True

## 72 CONSTRUCTING PIE CHARTS

**Exercise** 72A

**1**

| Favourite TV | Frequency | Angle |
|---|---|---|
| Soaps | 53 | 106° |
| Sport | 17 | 34° |
| Films | 40 | 80° |
| Comedy | 18 | 36° |
| Music | 26 | 52° |
| Drama | 26 | 52° |
| Total | 180 | 360° |

**Favourite TV**

**2**

| Pizza delivery | Frequency | Angle |
|---|---|---|
| 1–2 km | 43 | 164.7° |
| 2–5 km | 21 | 80.4° |
| 5–7 km | 17 | 65.1° |
| Over 7 km | 6 | 23.0° |
| Under 1 km | 7 | 26.8° |
| Total | 94 | 360° |

**Pizza delivery**

**3**

| Types of meal | Frequency | Angle |
|---|---|---|
| Beefburger | 25 | 128.6° |
| Vegetarian | 15 | 77.1° |
| Fish | 7 | 36.0° |
| Sausage | 13 | 66.9° |
| Roll | 10 | 51.4° |
| Total | 70 | 360° |

**Snacks**

**4**

| Transport to work | Frequency | Angle |
|---|---|---|
| Walk | 92 | 85.8° |
| Cycle | 75 | 69.9° |
| Car | 145 | 135.2° |
| Train | 20 | 18.7° |
| Coach or bus | 38 | 35.4° |
| Motor bike | 16 | 14.9° |
| Total | 386 | 360° |

**Transport to work**

**5**

| Tropical forests | Frequency | Angle |
|---|---|---|
| Asia | 16% | 57.6° |
| Africa | 35% | 126° |
| Latin America | 49% | 176.4° |
| Total | 100% | 360° |

WWF (1992)

**Tropical forests**

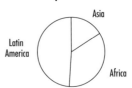

**6**

| Energy consumption 1987 | Per cent | Angle |
|---|---|---|
| USA | 24% | 86.4° |
| USSR | 17.5% | 63° |
| W.Europe | 17% | 61.2° |
| China | 9% | 32.4° |
| Japan | 5% | 18° |
| Africa | 2.5% | 9° |
| Rest of world | 25% | 90° |
| Total | 100 | 360° |

**World energy consumption**

**7**

| Rainfall | Rain (mm) | Angle |
|---|---|---|
| Jan–Mar | 74 | 16.6° |
| Apr–Jun | 481 | 108.0° |
| Jul–Sep | 910 | 204.2° |
| Oct–Dec | 139 | 31.2° |
| Total | 1604 | 360° |

**Rainfall**

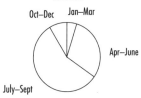

**8**

| Literacy – Women | Per cent | Angle |
|---|---|---|
| Can read | 54 | 194.4° |
| Cannot read | 46 | 165.6° |

| Literacy – Men | Per cent | Angle |
|---|---|---|
| Can read | 70 | 252° |
| Cannot read | 30 | 108° |

WWF (1992)

**Female literacy**

**Male literacy**

## Exercise 72B

**1**

| Best holiday | Frequency | Angle |
|---|---|---|
| Ski | 13 | 26° |
| Winter sun | 35 | 70° |
| Summer sun | 92 | 184° |
| Touring | 17 | 34° |
| UK | 23 | 46° |
| Total | 180 | 360° |

**Best holiday**

**2**

| Chef's favourite | Frequency | Angle |
|---|---|---|
| Main meal | 32 | 104.7° |
| Starters | 12 | 39.3° |
| Desserts | 24 | 78.5° |
| Specials | 42 | 137.5° |
| Total | 110 | 360° |

**Chef's favourites**

**3**

| Favourite event | Frequency | Angle |
|---|---|---|
| Long distance | 15 | 61.4° |
| Middle distance | 37 | 151.4° |
| Hurdles | 8 | 32.7° |
| Sprint | 28 | 114.5° |
| Total | 88 | 360° |

**Favourite event**

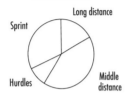

**4**

| UK toxic waste | Per cent | Angle |
|---|---|---|
| Dumping at sea | 8 | 28.8° |
| Treatment | 8 | 28.8° |
| Burning | 2 | 7.2° |
| Landfill | 82 | 295.2° |
| Total | 100 | 360° |

WWF UK's *Atlas of the Environment*

**UK toxic waste**

**5**

| World energy consumption | Per cent | Angle |
|---|---|---|
| Natural gas | 24 | 86.4° |
| Coal | 32 | 115.2° |
| Oil | 35 | 126° |
| Nuclear | 7 | 25.2° |
| Hydroelectric etc. | 2 | 7.2° |
| Total | 100 | 360° |

**World energy consumption**

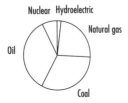

**6**

| Literacy (1990) | Per cent | Angle |
|---|---|---|
| Developing | 98 | 172.9° |
| Developed | 2 | 3.5° |
| Africa | 54 | 95.3° |
| Asia | 33 | 58.2° |
| Latin America | 17 | 30.0° |
| Total | 204 | 360° |

**Literacy**

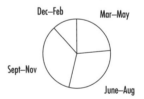

**7**

| Rainfall – Tokyo | Rain (mm) | Angle |
|---|---|---|
| Mar–May | 367 | 84.6° |
| Jun–Aug | 475 | 109.5° |
| Sep–Nov | 538 | 124.0° |
| Dec–Feb | 182 | 41.9° |
| Total | 1562 | 360° |

**Rainfall – Tokyo**

**8**

| Team results 1994 | Frequency | Angle |
|---|---|---|
| Win | 12 | 216° |
| Draw | 7 | 126° |
| Lose | 1 | 18° |
| Total | 20 | 360° |

| Team results 1995 | Frequency | Angle |
|---|---|---|
| Win | 7 | 126° |
| Draw | 8 | 144° |
| Lose | 5 | 90° |
| Total | 20 | 360° |

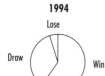

## 73 MEAN, MEDIAN, MODE AND RANGE FOR GROUPED DATA

### Exercise 73A

**1**

| Pledges, P(£) | Frequency, f | f × P |
|---|---|---|
| 1 | 39 | 39 |
| 2 | 15 | 30 |
| 3 | 5 | 15 |
| 4 | 3 | 12 |
| 5 | 8 | 40 |
| 6 | 1 | 6 |
| Totals | 71 | 142 |

(a) Mean = 2
(b) Median = £1
(c) Modal class: £1
(d) Range = 6 − 1 = £5

**2**

| No. of Heads, h | Frequency, f | f × h |
|---|---|---|
| 0 | 2 | 0 |
| 1 | 12 | 12 |
| 2 | 20 | 40 |
| 3 | 24 | 72 |
| 4 | 11 | 44 |
| 5 | 2 | 10 |
| Totals | 71 | 178 |

(a) Mean = 2.51 heads
(b) Median = 3 heads
(c) Modal class: 3 heads
(d) Range = 5 − 0 = 5 heads

**3**

| Mass (kg) | Frequency, f | Mid-value, m | f × m |
|---|---|---|---|
| $0 \leq M < 1$ | 2 | 0.5 | 1 |
| $1 \leq M < 2$ | 9 | 1.5 | 13.5 |
| $2 \leq M < 3$ | 11 | 2.5 | 27.5 |
| $3 \leq M < 4$ | 24 | 3.5 | 84 |
| $4 \leq M < 5$ | 4 | 4.5 | 18 |
| Total | 50 | | 144 |

(a) Mean = 2.88 kg
(b) Median = 3.5 kg
(c) Modal class: $3 \leq M < 4$ kg
(d) Max range = 5 − 0 = 5 kg

**4**

| Length (cm) | Frequency, f | Mid-value, m | f × m |
|---|---|---|---|
| $7 \leq L < 8$ | 35 | 7.5 | 262.5 |
| $8 \leq L < 9$ | 23 | 8.5 | 195.5 |
| $9 \leq L < 10$ | 11 | 9.5 | 104.5 |
| $10 \leq L < 11$ | 20 | 10.5 | 210 |
| $11 \leq L < 12$ | 5 | 11.5 | 57.5 |
| $12 \leq L < 13$ | 6 | 12.5 | 75 |
| Total | 100 | | 905 |

(a) Mean = 9.05 cm
(b) Median = 8.5 cm
(c) Modal class: $7 \leq L < 8$ cm
(d) Max. range = 13 − 7 = 6 cm

**5**

| Miles | Frequency, f | Mid-value, m | f × m |
|---|---|---|---|
| $10 \leq M < 20$ | 1 | 15 | 15 |
| $20 \leq M < 30$ | 2 | 25 | 50 |
| $30 \leq M < 40$ | 7 | 35 | 245 |
| $40 \leq M < 50$ | 45 | 45 | 2025 |
| $50 \leq M < 60$ | 22 | 55 | 1210 |
| $60 \leq M < 70$ | 2 | 65 | 130 |
| Total | 79 | | 3675 |

(a) Mean = 46.5
(b) Median = 45 km
(c) Modal class: $40 \leq M < 50$ km
(d) Max. range = 70 − 10 = 60 km

**6**

| Capacity (ml) | Frequency, f | Mid-value, m | f × m |
|---|---|---|---|
| $400 \leq C < 450$ | 3 | 425 | 1275 |
| $450 \leq C < 500$ | 3 | 475 | 1425 |
| $500 \leq C < 550$ | 13 | 525 | 6825 |
| $550 \leq C < 600$ | 5 | 575 | 2875 |
| $600 \leq C < 650$ | 6 | 625 | 3750 |
| $650 \leq C < 700$ | 3 | 675 | 2025 |
| Total | 33 | | 18175 |

(a) Mean = 551 ml
(b) Median = 525 ml
(c) Modal class: $500 \leq C < 550$ ml
(d) Max. range = 700 − 400 = 300 ml

**7**

| Weight (kg) | Frequency, $f$ | Mid-value, $m$ | $f \times m$ |
|---|---|---|---|
| $30 \leq w < 40$ | 5 | 35 | 175 |
| $40 \leq w < 50$ | 7 | 45 | 315 |
| $50 \leq w < 60$ | 58 | 55 | 3190 |
| $60 \leq w < 70$ | 6 | 65 | 390 |
| $70 \leq w < 80$ | 3 | 75 | 225 |
| $80 \leq w < 90$ | 1 | 85 | 85 |
| Total | 80 | | 4380 |

(a) Mean = 54.75 kg
(b) Median = 55 kg
(c) Modal class: $50 \leq w < 60$ kg
(d) Max. range = 90 − 30 = 60 kg

**8**

| Speed (m.p.h.) | Frequency, $f$ | Mid-value, $m$ | $f \times m$ |
|---|---|---|---|
| $0 \leq S < 10$ | 3 | 5 | 15 |
| $10 \leq S < 20$ | 4 | 15 | 60 |
| $20 \leq S < 30$ | 6 | 25 | 150 |
| $30 \leq S < 40$ | 12 | 35 | 420 |
| $40 \leq S < 50$ | 9 | 45 | 405 |
| $50 \leq S < 60$ | 6 | 55 | 330 |
| Total | 40 | | 1380 |

(a) Mean = 34.5 m.p.h.
(b) Median = 35 m.p.h.
(c) Modal class: $30 \leq S < 40$ m.p.h.
(d) Max. range = 60 − 0 = 60 m.p.h.

**Exercise** 73B

**1**

| Years, $y$ | Frequency, $f$ | $f \times y$ |
|---|---|---|
| 0 | 20 | 0 |
| 1 | 55 | 55 |
| 2 | 32 | 64 |
| 3 | 16 | 48 |
| 4 | 9 | 36 |
| Totals | 132 | 203 |

(a) Mean = 1.54 years
(b) Median = 1 year
(c) Modal class: 1 year
(d) Range = 4 − 0 = 4 years

**2**

| Score, $s$ | Frequency, $f$ | $f \times s$ |
|---|---|---|
| 1 | 15 | 15 |
| 2 | 38 | 76 |
| 3 | 14 | 42 |
| 4 | 15 | 60 |
| 5 | 5 | 25 |
| 6 | 13 | 78 |
| Totals | 100 | 296 |

(a) Mean = 2.96
(b) Median = 2
(c) Modal class: 2
(d) Range = 6 − 1 = £5

**3**

| Mass (t) | Frequency, $f$ | Mid-value, $m$ | $f \times m$ |
|---|---|---|---|
| $0 \leq M < 2$ | 66 | 1 | 66 |
| $2 \leq M < 4$ | 8 | 3 | 24 |
| $4 \leq M < 6$ | 4 | 5 | 20 |
| $6 \leq M < 8$ | 12 | 7 | 84 |
| $8 \leq M < 10$ | 7 | 9 | 63 |
| Total | 97 | | 257 |

(a) Mean = 2.65 t
(b) Median = 1 t
(c) Modal class: $0 \leq M < 2$ t
(d) Max. range = 10 − 0 = 10 t

**4**

| Ages (years) | Frequency, $f$ | Mid-value, $m$ | $f \times m$ |
|---|---|---|---|
| $15 \leq y < 20$ | 10 | 17.5 | 175 |
| $20 \leq y < 25$ | 8 | 22.5 | 180 |
| $25 \leq y < 30$ | 6 | 27.5 | 165 |
| $30 \leq y < 35$ | 7 | 32.5 | 227.5 |
| $35 \leq y < 40$ | 4 | 37.5 | 150 |
| $40 \leq y < 45$ | 5 | 42.5 | 212.5 |
| Total | 40 | | 1110 |

(a) Mean = 27.75 years
(b) Median = 27.5 years
(c) Modal class: $15 \leq y < 20$ years
(d) Max. range = 45 − 15 = 30 years

**5**

| Time (s) | Frequency, $f$ | Mid-value, $m$ | $f \times m$ |
|---|---|---|---|
| $0 \leq T < 5$ | 1 | 2.5 | 2.5 |
| $5 \leq T < 10$ | 10 | 7.5 | 75 |
| $10 \leq T < 15$ | 17 | 12.5 | 212.5 |
| $15 \leq T < 20$ | 11 | 17.5 | 192.5 |
| $20 \leq T < 25$ | 6 | 22.5 | 135 |
| $25 \leq T < 30$ | 4 | 27.5 | 110 |
| $30 \leq T < 35$ | 2 | 32.5 | 65 |
| Total | 51 | | 792.5 |

(a) Mean = 15.5 s
(b) Median = 12.5 s
(c) Modal class: $10 \leq T < 15$ s
(d) Max. range = 35 − 0 = 35 s

**6**

| Distance (paces) | Frequency, $f$ | Mid-value, $m$ | $f \times m$ |
|---|---|---|---|
| $30 \leq d < 40$ | 3 | 35 | 105 |
| $40 \leq d < 50$ | 5 | 45 | 225 |
| $50 \leq d < 60$ | 17 | 55 | 935 |
| $60 \leq d < 70$ | 9 | 65 | 585 |
| $70 \leq d < 80$ | 4 | 75 | 300 |
| $80 \leq d < 90$ | 3 | 85 | 255 |
| $90 \leq d < 100$ | 2 | 95 | 190 |
| Total | 43 | | 2595 |

(a) Mean = 60.3 paces
(b) Median = 55 paces
(c) Modal class: $50 \leq d < 60$ paces
(d) Max. range = 100 − 30 = 70 paces

**7**

| Yields (t) | Frequency, $f$ | Mid-value, $m$ | $f \times m$ |
|---|---|---|---|
| $0 \leq y < 2$ | 3 | 1 | 3 |
| $2 \leq y < 4$ | 9 | 3 | 27 |
| $4 \leq y < 6$ | 18 | 5 | 90 |
| $6 \leq y < 8$ | 10 | 7 | 70 |
| $8 \leq y < 10$ | 13 | 9 | 117 |
| $10 \leq y < 12$ | 11 | 11 | 121 |
| Total | 64 | | 428 |

(a) Mean = 6.69 t
(b) Median = 7 t
(c) Modal class: $4 \leq y < 6$ t
(d) Max. range = 12 − 0 = 12 t

**8**

| Height (cm) | Frequency, $f$ | Mid-value, $m$ | $f \times m$ |
|---|---|---|---|
| $140 \leq h < 150$ | 4 | 145 | 580 |
| $150 \leq h < 160$ | 13 | 155 | 2015 |
| $160 \leq h < 170$ | 7 | 165 | 1155 |
| $170 \leq h < 180$ | 4 | 175 | 700 |
| $180 \leq h < 190$ | 2 | 185 | 370 |
| $190 \leq h < 200$ | 1 | 195 | 195 |
| Total | 31 | | 5015 |

(a) Mean = 162 cm
(b) Median = 155 cm
(c) Modal class: $150 \leq h < 160$ cm
(d) Max. range = 200 − 140 = 60 cm

# REVISION

## Exercise G

**1**

| Time, $T$ (min) | Tally | Frequency |
|---|---|---|
| $15 \leq T < 20$ | III | 3 |
| $20 \leq T < 25$ | JHT JHT IIII | 14 |
| $25 \leq T < 30$ | JHT IIII | 9 |
| $30 \leq T < 35$ | JHT I | 6 |
| $35 \leq T < 40$ | IIII | 4 |
| $40 \leq T < 45$ | III | 3 |
| $45 \leq T < 50$ | I | 1 |

**2** (a) 11    (b) 12    (c) $\frac{3}{7}$    (d) 36%

**3**

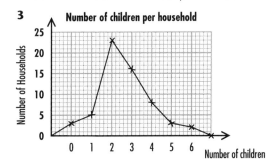

**4** (a) Angles: 18°, 30°, 138°, 96°, 48°, 18°, 12°
(b)

Number of children per household

**5** (a) 27.3    (b) 27    (c) 35
(d) $20 \leq T < 25$ min

## Exercise GG

**1** (a)

| Weight, w (kg) | Frequency |
|---|---|
| 1.970 ≤ w < 1.980 | 2 |
| 1.980 ≤ w < 1.990 | 7 |
| 1.990 ≤ w < 2.000 | 10 |
| 2.000 ≤ w < 2.010 | 12 |
| 2.010 ≤ w < 2.020 | 3 |
| 2.020 ≤ w < 2.030 | 2 |

(b)

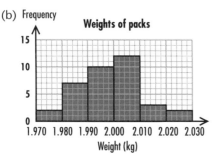

**Weights of packs**

**2** (a) 23    (b) 25    (c) 56    (d) 44.6%

**3** (a) 31

(b) 12 ≤ weight < 13 kg for each

(c) A: 9(.000) kg; B: 8(.000) kg

(d) Variety A is more consistent having most of the yields around 12–14 kg. Variety B has yields that are spread over the full range of weights.

**4** (a)

| Distance, d (m) | Frequency | Angle |
|---|---|---|
| 0 ≤ d < 25 | 3 | 13.3° |
| 25 ≤ d < 50 | 5 | 22.2° |
| 50 ≤ d < 75 | 8 | 35.6° |
| 75 ≤ d < 100 | 7 | 31.1° |
| 100 ≤ d < 125 | 9 | 40° |
| 125 ≤ d < 150 | 17 | 75.6° |
| 150 ≤ d < 175 | 32 | 142.2° |
| Totals | 81 | 360° |

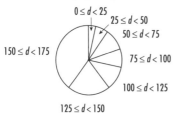

**Paper printing**

**5** (a) Total weight = 71.903 kg;
mean = 1.997 kg; median = 1.999 kg

(b)

| Weight, w (kg) | Frequency, f | Mid-interval, m | f × m |
|---|---|---|---|
| 1.970 ≤ w < 1.980 | 2 | 1.975 | 3.95 |
| 1.980 ≤ w < 1.990 | 7 | 1.985 | 13.895 |
| 1.990 ≤ w < 2.000 | 10 | 1.995 | 19.95 |
| 2.000 ≤ w < 2.010 | 12 | 2.005 | 24.06 |
| 2.010 ≤ w < 2.020 | 3 | 2.015 | 6.045 |
| 2.020 ≤ w < 2.030 | 2 | 2.025 | 4.05 |
| Totals | 36 | | 71.95 |

Mean = 1.999 kg
Median = 1.995 kg

(c) The estimates are good approximations in both cases.

(d) 2.000 ≤ w < 2.010 kg

(e) (i) Maximum range from grouped frequency table is 2.030 – 1.970 = 0.060 kg

    (ii) The actual range is
2.021 – 1.972 = 0.049

    (iii) The range for the grouped frequency table has to extend to the extremes of the first and last class intervals just in case there are terms with these values.

# 74 SCATTER DIAGRAMS: DRAWING FROM GIVEN DATA

## Exercise 74A

The diagrams also carry information for use in Exercise 75A.

**1**

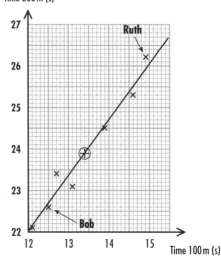

**2**

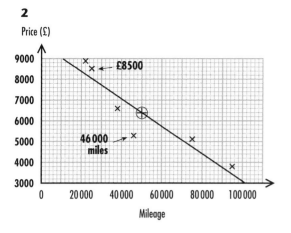

**3**

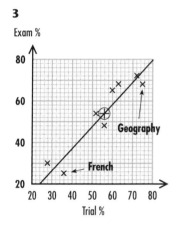

**4**

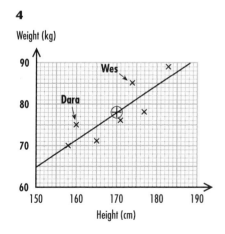

**5**

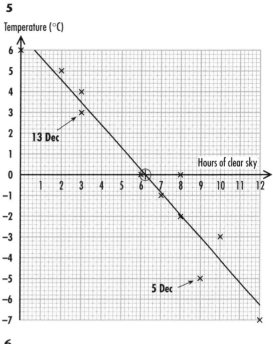

**6**

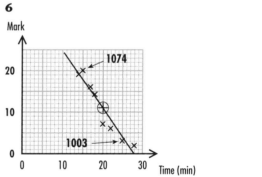

## Exercise 74B

The diagrams also carry information for use in Exercise 75B.

**1**

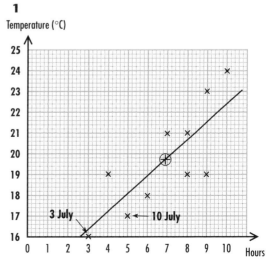

**2**

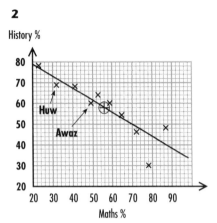

History %

**3**

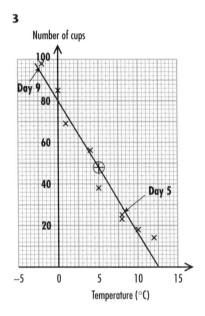

Number of cups

**4**

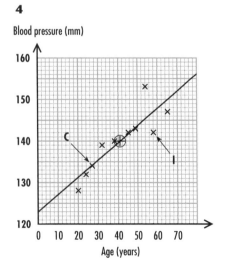

Blood pressure (mm)

**5**

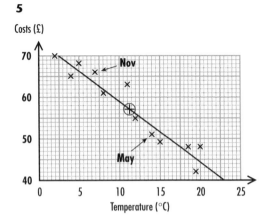

Costs (£)

**6**

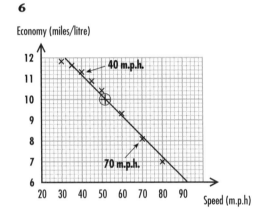

Economy (miles/litre)

# 75 SCATTER DIAGRAMS: LINE OF 'BEST FIT'

**Exercise** 75A

Refer to diagrams of answers to Exercise 74A. Answers will vary according to line of best fit drawn but the answers given are read from calculated lines of best fit.

**1** (a) (13.4, 23.9)
(d) (i) 26.1 s (200 m)  (ii) 12.7 s (100 m)

**2** (a) (50 200 miles, £6370)
(d) (i) £4400  (ii) 10 000 miles

**3** (a) (53%, 54%)
(d) (i) 80% Exam  (ii) 24% Trial

**4** (a) (170 cm, 77.7 kg)
(d) (i) 84 kg  (ii) 160 cm

**5** (a) (0°C, 6.2 h)
(d) (i) 10 h  (ii) 3.6°C

**6** (a) (19.9 min, 10.9)
(d) (i) 18  (ii) 28 min

## Exercise 75B

Refer to diagrams of answers to Exercise 74B. Answers will vary according to line of best fit drawn but the answers given are read from calculated lines of best fit.

**1** (a) (6.9 h, 19.7°C)
  (d) (i) 15 s (100 m)  (ii) 23 s (200 m)
**2** (a) (56%, 58%)     (d) (i) 70%  (ii) 45%
**3** (a) (5.1°, 47 cups)    (d) (i) 80 cups  (ii) 11°C
**4** (a) (41 years, 140 mm)
  (d) (i) 148 mm  (ii) 17 years
**5** (a) (11.25°C, £57.20) (d) (i) 15°C  (ii) £65
**6** (a) (51.3 m.p.h., 10.1 miles/litre)
  (d) (i) 32 m.p.h.  (ii) 6.2 miles per litre

# 76 SCATTER DIAGRAMS: CORRELATION

## Exercise 76A

**1** Positive            **2** Negative
**3** Perfect positive    **4** No correlation
**5** Perfect negative    **6** Negative
**7** Positive            **8** No correlation
**9** Perfect negative    **10** No correlation
**11** No correlation     **12** Perfect positive

## Exercise 76B

**1** Perfect negative    **2** Negative
**3** No correlation      **4** Perfect positive
**5** Perfect positive    **6** Negative
**7** Positive            **8** No correlation
**9** Perfect positive    **10** Perfect negative
**11** Negative           **12** No correlation

## Exercise 76C

**1** Positive (good)         **2** Negative (moderate)
**3** Positive (moderate)     **4** Positive (moderate)
**5** Negative (moderate)     **6** Negative (good)

## Exercise 76D

**1** Positive (moderate)     **2** Negative (moderate)
**3** Negative (good)         **4** Positive (moderate)
**5** Negative (moderate)     **6** Negative (good)

# 77 PROBABILITY: RELATIVE FREQUENCY USED TO MAKE ESTIMATES

## Exercise 77A

**1** Profit = £15
**2** 800
**3** (a) $\frac{2}{5}$ or 0.4   (b) $\frac{3}{50}$ or 0.06   (c) $\frac{7}{100}$ or 0.07

---

  (d) No – not exactly but with the same type of distribution
**4** (a) $\frac{2}{15}$ or 0.133        (b) $\frac{7}{8}$ or 0.875
  (c) $\frac{1}{60}$ or 0.0167          (d) No profit/ no loss
**5** (a) $\frac{3}{100}$ or 0.03        (b) $\frac{4}{5}$ or 0.8
  (c) 80                                (d) 2
**6** (a) 0.203                          (b) $\frac{203}{300}$ or 0.677
  (c) 7                                 (d) 29
**7** (a) 11%      (b) 19 000      (c) 17 000
  (d) Yes
**8** (a) 6          (b) $\frac{1}{20}$
  (c) Even; total for evens is 53 compared with 47 for odds.
  (d) A small loss, 50p

## Exercise 77B

**1** 540 g
**2** £135
**3** (a) Yes, 6.     (b) $\frac{4}{15}$ or 0.267     (c) $\frac{3}{5}$
  (d) 2:3
**4** (a) 5 or 6 – not 5.5        (b) $\frac{1}{4}$
  (c) 21%                         (d) 11
**5** (a) 0.16                    (b) $\frac{40}{43}$ or 0.930
  (c) $\frac{1}{26}$ or 0.0385
  (d) Less – because the probability is $\frac{6}{11}$ for the range 11–16°C and 11°C is on the low end of this range.
**6** (a) $\frac{25}{78}$ or 0.321     (b) $\frac{7}{12}$ or 0.583
  (c) $\frac{105}{156}$ or 0.673      (d) 29
**7** (a) 10 800                  (b) 7200
  (c) Same                        (d) Only Tony Smith
**8** (a) 3 heads
  (b) It is approximately symmetrical.
  (c) $\frac{3}{100}$
  (d) No – there is one more '6 heads' by chance.

# 78 PROBABILITY: COMPLEMENTARY EVENTS

## Exercise 78A

**1** (a) $\frac{4}{9}$      (b) Picking a consonant      (c) $\frac{5}{9}$
**2** (a) 0.429
  (b) Choosing a day that does not have six letters
  (c) 0.571
**3** (a) $\frac{2}{3}$      (b) Selecting a 10p coin      (c) $\frac{1}{3}$

**4** (a) 0.2    (b) Getting 4 or less    (c) 0.8

**5** (a) $\frac{6}{7}$    (b) Choosing a 2    (c) $\frac{1}{7}$

**6** (a) $\frac{1}{3}$    (b) Getting a score of 3 or more

   (c) $\frac{2}{3}$

**7** (a) 0.3    (b) Choosing a blue or green disc

   (c) 0.7

**8** (a) $\frac{5}{6}$    (b) Selecting a 5p coin    (c) $\frac{1}{6}$

**9** (a) 0.005    (b) Losing    (c) 0.995

**10** (a) $\frac{1}{3}$    (b) Choosing an odd digit    (c) $\frac{2}{3}$

**11** (a) 0    (b) Picking a day with a *y*    (c) 1

**12** (a) 0.625    (b) Choosing an even digit

   (c) 0.375

**13** (a) $\frac{3}{5}$    (b) Picking a vowel    (c) $\frac{2}{5}$

**14** (a) $\frac{1}{3}$

   (b) Choosing a month that ends in a letter

   other than R

   (c) $\frac{2}{3}$

**15** (a) $\frac{5}{8}$    (b) Choosing a red pencil    (c) $\frac{3}{8}$

**16** (a) $\frac{4}{9}$    (b) Selecting a 20p coin    (c) $\frac{5}{9}$

**17** (a) 0.0125    (b) Losing    (c) 0.9875

**18** (a) $\frac{2}{7}$

   (b) Choosing a shape that does not necessarily

   have a right angle

   (c) $\frac{5}{7}$

**19** (a) $\frac{5}{9}$    (b) Picking a vowel    (c) $\frac{4}{9}$

**20** (a) 0.75    (b) Getting less than 3    (c) 0.25

## Exercise 78B

**1** (a) $\frac{5}{6}$    (b) Getting a score of 1    (c) $\frac{1}{6}$

**2** (a) $\frac{3}{8}$    (b) Picking a consonant    (c) $\frac{5}{8}$

**3** (a) 0.125    (b) Choosing a nought

   (c) 0.875

**4** (a) $\frac{2}{7}$

   (b) Choosing a day that does not have eight

   letters

   (c) $\frac{5}{7}$

**5** (a) $\frac{5}{7}$    (b) Selecting a 5p coin    (c) $\frac{2}{7}$

**6** (a) 0.03    (b) Losing    (c) 0.97

**7** (a) $\frac{7}{12}$    (b) Choosing a blue or yellow disc

   (c) $\frac{5}{12}$

**8** (a) $\frac{2}{5}$    (b) Selecting a 20p coin    (c) $\frac{3}{5}$

**9** (a) $\frac{2}{5}$    (b) Picking a consonant    (c) $\frac{3}{5}$

**10** (a) $\frac{3}{7}$    (b) Choosing an even digit    (c) $\frac{4}{7}$

**11** (a) $\frac{2}{5}$    (b) Choosing a blue pencil    (c) $\frac{3}{5}$

**12** (a) $\frac{3}{5}$    (b) Getting more than 2    (c) $\frac{2}{5}$

**13** (a) 0.4    (b) Choosing an odd digit    (c) 0.6

**14** (a) 0    (b) Picking a 5p coin    (c) 1

**15** (a) $\frac{1}{3}$

   (b) Choosing a month that ends in a letter

   other than Y.

   (c) $\frac{2}{3}$

**16** (a) 0.25    (b) Picking a spade, club or heart

   (c) 0.75

**17** (a) $\frac{1}{2}$    (b) Picking a vowel    (c) $\frac{1}{2}$

**18** (a) $\frac{1}{3}$    (b) Selecting a 50p coin    (c) $\frac{2}{3}$

**19** (a) 0.002    (b) Losing    (c) 0.998

**20** (a) $\frac{7}{8}$    (b) Getting a 7    (c) $\frac{1}{8}$

# 79 PROBABILITY: TWO COMBINED, INDEPENDENT EVENTS

## Exercise 79A

**1**

| 1R | 1O | 1L | 1L | 1I | 1N | 1G |
|----|----|----|----|----|----|----|
| 2R | 2O | 2L | 2L | 2I | 2N | 2G |
| 3R | 3O | 3L | 3L | 3I | 3N | 3G |
| 4R | 4O | 4L | 4L | 4I | 4N | 4G |
| 5R | 5O | 5L | 5L | 5I | 5N | 5G |
| 6R | 6O | 6L | 6L | 6I | 6N | 6G |

(a) $\frac{1}{42}$    (b) $\frac{2}{21}$    (c) $\frac{1}{7}$    (d) 0

**2**

| 1R | 1B |
|----|----|
| 1R | 1B |
| 3R | 3B |
| 1R | 1B |
| 3R | 3B |

(a) 0.2    (b) 0.5    (c) 0    (d) 0.3

**3**

| GC | GD | GH | GS |
|----|----|----|----|
| GC | GD | GH | GS |
| GC | GD | GH | GS |
| GC | GD | GH | GS |
| YC | YD | YH | YS |
| YC | YD | YH | YS |
| YC | YD | YH | YS |

(a) $\frac{3}{28}$    (b) $\frac{2}{7}$    (c) $\frac{19}{28}$    (d) $\frac{5}{7}$

**4**

| wC | wO | wM | wM | wI | wT | wT | wE | wE |
|----|----|----|----|----|----|----|----|----|
| wC | wO | wM | wM | wI | wT | wT | wE | wE |
| bC | bO | bM | bM | bI | bT | bT | bE | bE |
| bC | bO | bM | bM | bI | bT | bT | bE | bE |
| bC | bO | bM | bM | bI | bT | bT | bE | bE |
| bC | bO | bM | bM | bI | bT | bT | bE | bE |

(a) 0.0741  (b) 0.0741  (c) 0.148  (d) 0.148

**5**

| C1 | C2 | C3 | C4 | C5 |
|----|----|----|----|----|
| D1 | D2 | D3 | D4 | D5 |
| H1 | H2 | H3 | H4 | H5 |
| S1 | S2 | S3 | S4 | S5 |

(a) 0.05  (b) 0.2  (c) 0.55  (d) 0.15

**6**

| $1+20p$ | $2+20p$ | $3+20p$ | $4+20p$ | $5+20p$ | $6+20p$ |
|---------|---------|---------|---------|---------|---------|
| $1+20p$ | $2+20p$ | $3+20p$ | $4+20p$ | $5+20p$ | $6+20p$ |
| $1+20p$ | $2+20p$ | $3+20p$ | $4+20p$ | $5+20p$ | $6+20p$ |
| $1+20p$ | $2+20p$ | $3+20p$ | $4+20p$ | $5+20p$ | $6+20p$ |
| $1+20p$ | $2+20p$ | $3+20p$ | $4+20p$ | $5+20p$ | $6+20p$ |
| $1+50p$ | $2+50p$ | $3+50p$ | $4+50p$ | $5+50p$ | $6+50p$ |
| $1+50p$ | $2+50p$ | $3+50p$ | $4+50p$ | $5+50p$ | $6+50p$ |

(a) $\frac{5}{42}$  (b) $\frac{1}{7}$  (c) $\frac{17}{42}$  (d) $\frac{5}{14}$

**7**

| Tb | Eb | Eb | Tb | Hb |
|----|----|----|----|----|
| Tr | Er | Er | Tr | Hr |

(a) 0.2  (b) 0.1  (c) 0.7  (d) 0

**8**

| RR | RR | RB | RB | RB | RB | RB |
|----|----|----|----|----|----|----|
| RR | RR | RB | RB | RB | RB | RB |
| BR | BR | BB | BB | BB | BB | BB |
| BR | BR | BB | BB | BB | BB | BB |
| BR | BR | BB | BB | BB | BB | BB |
| BR | BR | BB | BB | BB | BB | BB |
| BR | BR | BB | BB | BB | BB | BB |

(a) $\frac{4}{49}$  (b) $\frac{25}{49}$  (c) 1  (d) $\frac{20}{49}$

**9**

| 26 | 27 | 28 | 29 |
|----|----|----|----|
| 56 | 57 | 58 | 59 |
| 06 | 07 | 08 | 09 |
| 56 | 57 | 58 | 59 |
| 06 | 07 | 08 | 09 |

(a) 0.2  (b) 0.1  (c) 0.3  (d) 0.4

**10**

| B1 | B2 | B4 | B3 | B2 |
|----|----|----|----|----|
| R1 | R2 | R4 | R3 | R2 |

(a) $\frac{1}{10}$  (b) $\frac{3}{10}$  (c) $\frac{3}{10}$  (d) 0

## Exercise 79B

**1**

| 0R | 1R | 2R | 3R |
|----|----|----|----|
| 0R | 1R | 2R | 3R |
| 0R | 1R | 2R | 3R |
| 0B | 1B | 2B | 3B |
| 0B | 1B | 2B | 3B |

(a) 0.1  (b) 0.15  (c) 0.2  (d) 0.6

**2**

| C1 | C2 | C3 | C4 | C5 | C6 |
|----|----|----|----|----|----|
| H1 | H2 | H3 | H4 | H5 | H6 |
| O1 | O2 | O3 | O4 | O5 | O6 |
| O1 | O2 | O3 | O4 | O5 | O6 |
| S1 | S2 | S3 | S4 | S5 | S6 |
| E1 | E2 | E3 | E4 | E5 | E6 |

(a) $\frac{1}{36}$  (b) $\frac{1}{18}$  (c) $\frac{1}{9}$  (d) $\frac{1}{12}$

**3**

| R4 | R4 | R1 | R5 | R5 | R4 |
|----|----|----|----|----|----|
| B4 | B4 | B1 | B5 | B5 | B4 |

(a) $\frac{1}{12}$  (b) $\frac{1}{12}$  (c) $\frac{2}{3}$  (d) $\frac{1}{3}$

**4**

| R1 | R2 | R3 | R4 | R5 | R6 |
|----|----|----|----|----|----|
| R1 | R2 | R3 | R4 | R5 | R6 |
| R1 | R2 | R3 | R4 | R5 | R6 |
| R1 | R2 | R3 | R4 | R5 | R6 |
| R1 | R2 | R3 | R4 | R5 | R6 |
| B1 | B2 | B3 | B4 | B5 | B6 |
| B1 | B2 | B3 | B4 | B5 | B6 |
| B1 | B2 | B3 | B4 | B5 | B6 |

(a) $\frac{5}{48}$  (b) $\frac{11}{16}$  (c) $\frac{1}{4}$  (d) $\frac{11}{16}$

**5**

| CC | CD | CH | CS |
|----|----|----|----|
| DC | DD | DH | DS |
| HC | HD | HH | HS |
| SC | SD | SH | SS |

(a) 0.25  (b) 0.75  (c) 0.438  (d) 0.125

**6**

| 71 | 72 | 73 | 74 | 75 | 76 |
|----|----|----|----|----|----|
| 81 | 82 | 83 | 84 | 85 | 86 |
| 91 | 92 | 93 | 94 | 95 | 96 |
| 91 | 92 | 93 | 94 | 95 | 96 |
| 91 | 92 | 93 | 94 | 95 | 96 |

(a) 0.1  (b) 0.43  (c) 0.8  (d) 0

**7**

| 21 | 22 | 23 | 24 | 25 |
|----|----|----|----|----|
| 01 | 02 | 03 | 04 | 05 |
| 01 | 02 | 03 | 04 | 05 |
| 01 | 02 | 03 | 04 | 05 |
| 01 | 02 | 03 | 04 | 05 |
| 11 | 12 | 13 | 14 | 15 |

(a) $\frac{1}{10}$  (b) $\frac{1}{30}$  (c) $\frac{4}{15}$  (d) $\frac{11}{15}$

**8**

| sD | sA | sG | sG | sE | sR |
|----|----|----|----|----|----|
| cD | cA | cG | cG | cE | cR |
| dD | dA | dG | dG | dE | dR |
| hD | hA | hG | hG | hE | hR |

(a) 0.0417  (b) 0.0833  (c) 0.5  (d) 0.0833

**9**

20p + 20p  20p + 20p  20p + 20p  20p + 20p  20p + 20p
                              20p + 50p  20p + 50p
20p + 20p  20p + 20p  20p + 20p  20p + 20p  20p + 20p
                              20p + 50p  20p + 50p
20p + 20p  20p + 20p  20p + 20p  20p + 20p  20p + 20p
                              20p + 50p  20p + 50p
20p + 20p  20p + 20p  20p + 20p  20p + 20p  20p + 20p
                              20p + 50p  20p + 50p
20p + 20p  20p + 20p  20p + 20p  20p + 20p  20p + 20p
                              20p + 50p  20p + 50p
20p + 20p  20p + 20p  20p + 20p  20p + 20p  20p + 20p
                              20p + 50p  20p + 50p
50p + 20p  50p + 20p  50p + 20p  50p + 20p  50p + 20p
                              50p + 50p  50p + 50p
50p + 20p  50p + 20p  50p + 20p  50p + 20p  50p + 20p
                              50p + 50p  50p + 50p

(a) $\frac{20}{49}$    (b) $\frac{4}{49}$    (c) $\frac{25}{49}$    (d) $\frac{29}{49}$

**10**  Cc    Cd    Ch    Cs
      Dc    Dd    Dh    Ds
      Hc    Hd    Hh    Hs
      Sc    Sd    Sh    Ss

(a) 0.0625    (b) 0.125    (c) 0.25    (d) 0.75

## $R$ EVISION

---

### Exercise $H$

**1** (a) Science 46%, English 52%
   (b) Science 35%, English 60%
   (c) 32%    (d) 20%    (e) Negative

**2** (a) Positive    (b) No correlation
   (c) Negative    (d) Perfect positive

**3** (a) Yes, C and D occur more often
   (b) 14 or 15    (c) 0.39
   (d) 10 or 11 (not 10.5)

**4** (a) (i) $\frac{3}{7}$ (ii) Picking a consonant  (iii) $\frac{4}{7}$
   (b) (i) $\frac{1}{400}$ (ii) Losing  (iii) $\frac{399}{400}$
   (c) (i) $\frac{3}{5}$ (ii) Getting 4 or more  (iii) $\frac{2}{5}$

**5** (a) 1, 1    1, 2    1, 3    1, 4    1, 5    1, 6
      2, 1    2, 2    2, 3    2, 4    2, 5    2, 6
      3, 1    3, 2    3, 3    3, 4    3, 5    3, 6
      4, 1    4, 2    4, 3    4, 4    4, 5    4, 6
      5, 1    5, 2    5, 3    5, 4    5, 5    5, 6
      6, 1    6, 2    6, 3    6, 4    6, 5    6, 6

(b)

| Total score | 2 | 3 | 4 | 5 | 6 | 7 | 8 | 9 | 10 | 11 | 12 |
|---|---|---|---|---|---|---|---|---|---|---|---|
| Frequency | 1 | 2 | 3 | 4 | 5 | 6 | 5 | 4 | 3 | 2 | 1 |

(c)

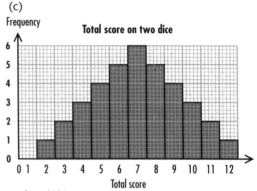

**Total score on two dice**

### Exercise $HH$

**1** (a) 6.7 h, 20.6°C
   (b) and (c)

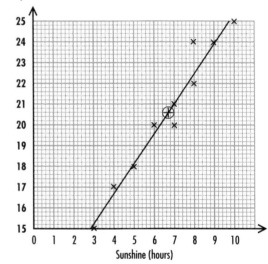

(d) (i) 22.5°C  (ii) 5.6 h    (e) Positive

**2** (a) 12 150    (b) 1350    (c) 12 600
   (d) Any of the three

**3** (a) rr    rr    rr    rb    rb    rb    rb    rb
      rr    rr    rr    rb    rb    rb    rb    rb
      rr    rr    rr    rb    rb    rb    rb    rb
      br    br    br    bb    bb    bb    bb    bb
      br    br    br    bb    bb    bb    bb    bb
      br    br    br    bb    bb    bb    bb    bb
      br    br    br    bb    bb    bb    bb    bb
      br    br    br    bb    bb    bb    bb    bb

(b) (i) $\frac{9}{64}$ (ii) $\frac{25}{64}$ (iii) $\frac{17}{32}$ (iv) $\frac{15}{32}$ (v) $\frac{9}{64}$ (vi) $\frac{17}{32}$

(c) rr    rr    rb    rb    rb    rb    rb

(d) (i) $\frac{2}{7}$ (ii) $\frac{5}{7}$